Fundamental Chemistry

*A Revision Book for
First Certificate*

Fundamental Chemistry

A Revision Book for
First Certificate

M. J. Long and **J. H. White**

Heinemann Educational Books
London

Heinemann Educational Books

LONDON EDINBURGH MELBOURNE AUCKLAND TORONTO
HONG KONG SINGAPORE KUALA LUMPUR NEW DELHI
IBADAN NAIROBI JOHANNESBURG LUSAKA KINGSTON

ISBN 0 435 64520 X

Published by Heinemann Educational Books Ltd
48 Charles Street, London W1X 8AH

Printed in Great Britain by
Butler & Tanner Ltd, Frome and London

Preface

This book summarizes the knowledge that the authors feel pupils should have before they commence a formal sixth-form course in chemistry. The work is based throughout on the Periodic Classification with constant reference to the Activity Series.

To assist its value as a revision text, each Section has been written so that it can be read as a separate entity – aided by adequate cross-referencing. This means that teachers preparing pupils for the various examinations, or for no examination, can be selective in the Sections that they require their pupils to read. The pupils themselves can also quickly revise any topics about which they feel they have inadequate knowledge.

Modern nomenclature has been adopted in all cases where the pupils are able to see the relevance and logic of the system according to the chemical knowledge that they have. Modern units have also been used throughout except that 'litre' and 'dm^3' both appear. This should cause no confusion and it would seem to be unwise to abandon the litre just as it is likely to become common in everyday life.

1977
M. J. L.
J. H. W.

Contents

1   Elements

Several Greek philosophers had the idea that all the many and varied things around them might be made from one simple material. Thales (*c.* 600 B.C) thought it was water; Anaximander (a little later) thought it was fire. Aristotle (*c.* 350 B.C.) allowed four of these 'elementary' substances: earth, air, fire, and water. (Not one of these is an element in the modern sense.) To the Greeks a log of wood when heated might give out air and water, leaving ash (earth); any flame was the fire escaping. The alchemists added mercury, sulphur, and salt to the list of 'elementary' substances. Mercury was volatile (easily changed to vapour) and gave a metallic appearance. Sulphur was the 'inflammable principle'. Metals were supposed to consist of mercury and sulphur in various proportions. Salt was the 'principle of permanence' – quite unaffected by strong heating.

Robert Boyle, in his well-known book of 1661, *The Sceptical Chymist*, pointed out the absurdity of all this and his definition of elements was 'primitive and simple bodies of which the mixed ones are said to be composed and into which they are ultimately resolved'. Lavoisier, over a hundred years later, succeeded in getting rid of the Aristotelian and alchemical 'elements'. To Lavoisier an element was a fundamental substance that, as far as he knew, could not be split into simpler substances. In 1789, he listed 33 of these and, of the 33, 23 can be found in a modern list of elements. Aristotelian ideas, however, lasted for over 2000 years and even today the word 'elements' is used when talking about the weather: the wind (air), rain (water), and lightning (fire).

The discovery of atoms and sub-atomic particles (Section 2) has caused chemists to give a more precise definition of an element:

A substance that consists of atoms of one type and which cannot be further decomposed by chemical means.

During the nineteenth century many more elements were discovered, and by the beginning of the twentieth century the number of these fundamental substances had risen to nearly 90. These are the 'bricks' that the chemist uses to make the vast number of compounds and mixtures known today. The maximum number of 'natural' elements is 92. Atomic physicists have extended the number to 106 by the synthesis of new atoms but these are unstable and are not encountered in everyday life.

Of the 92 elements that may occur in nature, only about 40 are fairly common. About three-quarters of the 92 are classed as metals. Only two, at ordinary temperatures and pressures, are liquid – mercury and bromine – and only eleven are gases – hydrogen, nitrogen, oxygen, fluorine, chlorine, and six 'noble' gases (Section 24).

Relative abundance of the elements in nature

Table 1 shows the ten most abundant elements (by mass) found in nature (either free or in combination) and this includes the atmosphere, the hydrosphere (seas), and the Earth's crust to a depth of about 16 km.

Table 1

Element	Occurrence	Approx. percentage
1. Oxygen	Air, water, oxides, carbonates, sand (silica), clay, and other silicates	47
2. Silicon	Sand, clay, and other silicates	27
3. Aluminium	Clay, bauxite, cryolite. Most abundant metal	7
4. Iron	Oxide, hydroxide, carbonate, sulphide	5
5. Calcium	Chalk, limestone, marble, gypsum, anhydrite	3
6. Sodium	Chloride (common salt), sulphate, carbonate, nitrate	2.5
7. Potassium	Carnallite and some silicates	2.5
8. Magnesium	Carbonate, sulphate, chloride (in sea water) and some silicates	2
9. Hydrogen	Water, ammonium compounds, and many organic compounds	0.2
10. Titanium	Often associated with iron. Widespread in small quantities	0.1

2   Atoms

During the fifth and fourth centuries B.C., the Greeks decided that matter was composed of 'atoms'. The gradual wearing away of statues exposed to the atmosphere and other observations led them to the belief in particles that were too small to see. The Greeks believed that if an object could be cut into two parts, then four, eight, sixteen, etc., a limit would eventually be reached when the particle could not be further divided. To such a particle the name 'atom' was given – which means 'indivisible'.

Early in the nineteenth century, John Dalton, in his *New System of Chemical Philosophy* (1808), applied the theory of atoms to explain chemical combination and gave it a quantitative basis by attempting to find the relative weights of the atoms. Dalton assumed that all the atoms of one element are exactly alike, but different from the atoms of any other element. He pictured them as minute, hard, solid, and round. He gave them symbols, each one circular, thus:

oxygen

hydrogen

nitrogen

carbon

Dalton took as his standard element the lightest – hydrogen. If he found the weight of oxygen that would combine with one part by weight of hydrogen to form water and assumed that one atom of oxygen combined with one of hydrogen, then the weight of oxygen would represent its atomic weight or what is now, more correctly, called the **relative atomic mass.** Dalton's formula for water was

and his atomic weight for oxygen was 6.5. As we now know that *two* atoms of hydrogen combine with one of oxygen when water is formed and the formula of water is H_2O, with the atomic weight of oxygen, 16, Dalton's value should have been 8. However, as accurate quantitative work had scarcely been attempted before Lavoisier's time and apparatus was very crude, Dalton's work was surprisingly good. Molecules were not clearly recognized by Dalton who spoke of 'compound atoms'. Some of his formulae were correct as, for example, the oxides of carbon:

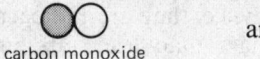 and

carbon monoxide   carbon dioxide

Dalton assumed the truth of the Law of Fixed Proportions (Section 78) and he established the Law of Multiple Proportions (Section 78). His pictorial symbols were soon replaced by the ones we use today – suggested by the Swedish chemist, Berzelius, in 1819. The initial (capital) letter of the element's name is used to represent one atom of that element. Where two or more elements have the same initial letter, a second (small) distinguishing letter is added thus: Carbon, **C**; Calcium, **Ca**; Cobalt, **Co**. Occasionally, the Latin name is used: Copper (cuprum), **Cu**; Iron (ferrum), **Fe**; silver (argentum), **Ag**; Mercury (hydrargyrum), **Hg**; Sodium (from the Latin word for soda – natrium), **Na**; Potassium (potash – kalium), **K**; Tin (stannum), **Sn**; Gold (aurum), **Au**.

During the twentieth century ideas on the nature of the atom have changed completely. From the work of the Curies on radioactivity followed by the researches of J. J. Thomson, Rutherford, and Aston, scientists have become convinced that the atom is *not* the smallest fundamental particle and that it is *not* solid. Physicists have now discovered many particles smaller than an atom, but we shall only consider the three most important: the **proton**, the **neutron**, and the **electron**. These are the ones most concerned with chemical activity.

Our current picture of the atom is that it is mostly space and the space occupied by the atom is known as the **atomic volume.** At the heart of this space is the incredibly small **nucleus** of the atom. In the rest of the space are the electrons. All atoms are neutral and have at least one electron in their atomic volumes. The nuclei of all atoms, except that of hydrogen, contain both protons and neutrons. The hydrogen nucleus has one proton and nothing else.

A **proton** is a unit of mass with one unit positive charge.

A **neutron** is a unit of the same mass but with no charge.

Outside the nucleus there are one or more extranuclear **electrons** occupying **energy levels** (or **shells**). The electron has some of the properties of a particle and some of a wave. It has a measurable mass and a unit negative charge. This mass is about 1/1840 that of a hydrogen atom – corresponding to about 9×10^{-28} g.

An atom is neutral and this means that the number of positive charges (protons) must always be equal to the number of electrons (which possess unit negative charges).

Almost all the mass of an atom is in the nucleus, so that an atom could lose one or more electrons without any significant change in its mass. How a number of mutually repulsive positive charges can exist closely packed in the nucleus without disruption is a difficult matter to explain. Everyone, nowadays, knows what vast stores of energy can be released when the nucleus is disrupted.

Consider an atom of calcium. Its relative atomic mass is 40. Its nucleus consists of 20 protons and 20 neutrons. Its atomic volume contains 20 electrons. The atom is neutral because the 20 positive charges (protons) are exactly balanced by the 20 negative charges (electrons). For all the simpler atoms (other than hydrogen) the number of protons in the nucleus is much the same as the number of neutrons. For elements of high relative atomic mass, however, the number of neutrons far exceeds the number of protons. Such atoms are unstable and break down spontaneously, i.e. they are radioactive. All atoms that have a greater relative atomic mass than lead, 207, are radioactive. Radium, for example, has a relative atomic mass of 226. Its atom has 88 extra-nuclear electrons and also, therefore, 88 protons. The number of neutrons in this case is much higher – 138.

Electrons have no fixed positions and do not move in defined orbits. Their positions at any moment are based on probability. In a relatively large atom the electrons appear in the various energy levels according to a definite plan. They do not, as was formerly believed, circle the nucleus like planets going round the Sun and, indeed, it is now known that not more than two electrons can occupy one **orbital.** (An orbital can only be defined in advanced mathematical terms. For the present purpose it may be thought of as the region in which an electron or, at most, two electrons are likely to be found.) Where there are eight electrons in an energy level, there will be four orbitals with two electrons in each.

The energy levels (or 'shells') are numbered outwards from the nucleus: 1, 2, 3, 4 (though previously they were lettered K, L. M, N.) The maximum number of electrons that each energy level can contain is given by the series:

$$2 \times 1^2, 2 \times 2^2, 2 \times 3^2, 2 \times 4^2$$

Energy level (shell) 1 (K) can therefore contain a maximum of 2 electrons
Energy level (shell) 2 (L) can therefore contain a maximum of 8 electrons
Energy level (shell) 3 (M) can therefore contain a maximum of 18 electrons
Energy level (shell) 4 (N) can therefore contain a maximum of 32 electrons

Consider the atom of sulphur of relative atomic mass 32. Its nucleus contains 16 protons and 16 neutrons. It has 16 electrons, which will be arranged:

Shell 1   2 electrons
Shell 2   8 electrons (four orbitals with 2 electrons in each)
Shell 3   6 electrons

Here the third shell is not completely filled.

Dalton chose hydrogen as the standard for the determination of relative

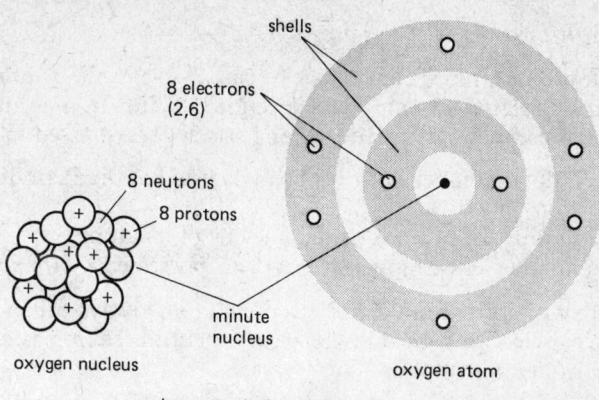

shells

8 electrons
(2,6)

8 neutrons

8 protons

minute
nucleus

oxygen nucleus

oxygen atom

electrons: shell 1 2 electrons
shell 2 6 electrons

Figure 1 A representation of the fairly simple oxygen atom (relative atomic mass, 16; number of protons, 8; number of neutrons, 8; number of electrons, 8)

atomic masses. It was soon found, however, that hydrogen was *not* a good choice because hydrogen is not very reactive and few other elements combine with it easily. Stas, about the middle of the nineteenth century, suggested that oxygen, with its relative atomic mass fixed as 16.00, would be a better standard. The suggestion was adopted and the system lasted until 1960/61 when scientists, by international agreement, changed the standard to carbon 12. The reason for this will become clear later in this section.

During the nineteenth century many chemists devoted their energies to determining, with ever increasing accuracy, the relative atomic masses of the elements then known and, by the end of the century, most of the values were known to four significant figures. When the standard element became oxygen 16.00, hydrogen correspondingly became 1.008 because it was found that 1 g of hydrogen actually combines with 7.94 g oxygen when water is formed. Knowing that the formula of water was H_2O this gave a value for oxygen (compared with hydrogen 1) of 15.88. When oxygen was fixed as exactly 16.00, values for the other elements were adjusted accordingly. The adoption of the carbon standard makes no appreciable difference to relative atomic masses when given to four significant figures. Today, **relative atomic mass** may be defined as follows.

Relative atomic mass is twelve times the ratio of the mass of an atom of the element to that of a carbon-12 atom.

Nearly all the values for relative atomic mass approximate to whole numbers and, noting this, Prout, as early as 1815, suggested that they ought all to be whole numbers. He thought that the atoms of other elements might be formed by compacting together various numbers of hydrogen atoms. The theory was abandoned when careful experiments fixed the relative atomic mass of chlorine as 35.45. As the atom at that time was still believed to be indivisible, 0.45 of an atom could not be accepted. We now know that Prout's theory was not far from the truth.

Atomic number

Chemists now find it more convenient to classify elements by **atomic number** rather than by relative atomic mass (Section 5). The atomic number of an element is represented by the letter Z and it can be expressed in three ways:

(1) It is the ordinal number of the element as listed in the Periodic Classification (Section 5).
(2) It is the number of protons in the nucleus of its atom.
(3) It is the number of extranuclear electrons possessed by the atom.

The number of protons in the nucleus of an atom always equals the number of extranuclear electrons because the atom is neutral. Taking as examples the atoms of calcium and sulphur:

Z for calcium is 20 because it is the twentieth element in the Periodic Classification; the nucleus of its atom has 20 protons and in the atomic volume there are 20 extranuclear electrons.

Z for sulphur is 16 because it is the sixteenth element in the Periodic Classification; the nucleus of its atom has 16 protons and in the atomic volume there are 16 extranuclear electrons.

It is perhaps worth remembering that for the first 20 elements of the Periodic Classification the atomic number is about half the relative atomic mass.

As it is now known that there are many particles smaller than an atom ('sub-atomic' particles) the atom is now defined as follows.

An atom is the smallest neutral particle of an element that has the chemical properties of that element and can take part in chemical change.

Isotopes

Following the work of Soddy, in 1914, it came to be accepted that relative atomic masses *are* whole numbers. The reason why chlorine has a relative atomic mass of 35.45 is because natural chlorine gas has atoms of two different masses 35 and 37 – these figures being known as **mass numbers**. When the relative atomic mass of chlorine is determined, the resulting figure is the average mass of a mixture of atoms 35 and 37. As the average is 35.45, the ratio of 35s to 37s must be about 3 : 1. The two chlorines have the same atomic number because they have the same number (17) of protons and electrons, but chlorine 37 has two extra neutrons in the nucleus of its atom. The chemical properties of the two are almost identical. Soddy gave them the name **isotopes** because both have the same place ('topos' is the Greek word for place) in the Periodic Classification (Section 5).

Isotopes of an element have the same atomic number but differ in atomic mass.

Nearly all the elements have two or more isotopes but often the atoms of mass differing from the principal isotope are in very small quantity. Oxygen-16 has also 0.20% oxygen-18 as well as 0.04% oxygen-17. Carbon has 98.89% carbon-12 and 1.11% carbon-13. Carbon-12 (also written ^{12}C) is now taken as the standard for atomic masses.

The isotopes of chlorine can be separated in several different ways. One of the earliest methods was that of diffusion. Molecules formed from chlorine-35 diffuse faster than those formed from chlorine-37 by the ratio $\frac{\sqrt{37}}{\sqrt{35}} = 1.03:1$ (Graham's Law Sections 13 and 78). This is not a large difference, but if chlorine gas is made to diffuse through a long porous plug, the 35 isotope will gradually outstrip the 37 isotope and the first gas to appear at the end of the plug will consist almost entirely of isotope 35. In recent years an apparatus has been devised that can separate isotopes directly and also record their relative masses. It is called a mass spectrometer and, with its aid, it has been possible to show that nearly every element has two or more isotopes. Aluminium, arsenic, cobalt, fluorine, iodine, manganese, and phosphorus are elements that seem to have no isotopes.

Round the symbol for the atom of an element, which is here represented as X, there are four places, *a,b,c,d*, where figures may be placed as shown:

At *a*, the *mass number* is given, at *b*, the *atomic number*, at *c*, any *ionic charge* (Section 8), and at *d* the *atomicity* (Section 3).

Chemists can thus easily distinguish between the two isotopes of chlorine by writing them as $^{37}_{17}Cl$ and $^{35}_{17}Cl$.

3 Molecules. The mole concept

MOLECULES

A molecule of an element or compound is the smallest particle that can exist in the free state whilst still having all the properties of that element or compound.

It is made up of one or more atoms. If a molecule has only one atom it is said to be monatomic and the only gaseous elements that are monatomic are the noble gases (Section 24). Some other elements, such as iodine, I, and mercury, Hg, can give monatomic vapours at higher temperatures. Most common gaseous elements are diatomic, e.g. nitrogen, N_2, oxygen, O_2, chlorine, Cl_2. Examples of triatomic molecules are ozone, O_3, water, H_2O, and carbon dioxide, CO_2. Ammonia, NH_3, is a tetratomic molecule.

The number of atoms in the molecule of an element or compound is known as the atomicity of the molecule.

Compounds that possess separate individual molecules are mainly covalent (Section 9) and these are usually gaseous or liquid. Most inorganic solid compounds consist of lattices of ions (Section 8) and have no separate molecules. If one considers a common organic compound such as ethanol, C_2H_5OH, and adds together the relative atomic masses, the total, 46, represents the **relative molecular mass** (formerly called 'molecular weight'). In the case of an ionic compound such as sodium chloride, the formula NaCl is the very simplest possible and only indicates that there is, in the crystal, one chloride ion for every sodium ion. As there are *no* separate molecules in the crystal structure, the sum of the relative atomic masses, 58.5, should be referred to as a **formula mass** and not a molecular mass.

Molecules that consist of a very large number of smaller ones (e.g. polymers, Section 57) are known as **macromolecules** (from the Greek 'makros' meaning 'large') or **'giant'** molecules. These terms are also sometimes used to describe crystals that contain large numbers of atoms (e.g. diamond, Section 49 or ions, Section 8).

In 1873, the Dutch scientist, van der Waals, explained deviations from Boyle's Law (Section 13) when gases are under high pressure by assuming that there must be weak attractive forces between molecules when the molecules are very close together. These weak attractive forces are negligible when the distance between the molecules increases. The explanation of the structure of graphite (Section 49) involves van der Waals' forces and in a solid organic compound, such as paraffin wax, it is probably these same forces that keep the molecules together. A very little energy, given by gentle heating, will overcome the attractive forces and cause the solid paraffin to melt (m.p. about 45 °C).

The Italian scientist, Avogadro, suggested as long ago as 1811, that

Equal volumes of gases, under the same conditions of temperature and pressure, contain equal numbers of molecules.

This was known as Avogadro's Hypothesis, but we are now so certain of its truth that it is frequently referred to as Avogadro's Law. Unfortunately, the hypothesis was not accepted until another Italian scientist, Cannizzaro, showed in 1858 that it gave the answer to many important problems. Cannizzaro was able to establish the formulae for the molecules of many gases and vapours including that for water, H_2O (see later in this section).

The volume occupied by the molar mass of any gas at s.t.p. is the same for all gases and this volume, 22.4 litres, is known as the molar volume.

('Molar mass' is the relative molecular mass expressed in grams and s.t.p. represents standard temperature (0 °C) and pressure (760 mm of mercury).)

In recent times, scientists have determined *the approximate number of molecules that are present in the molar volume and this figure, 6.0225×10^{23}, is known as the Avogadro Constant (symbol L) or Avogadro Number.* Avogadro himself probably never dreamed that such a number could be estimated. To obtain some idea of what a volume of 22.4 litres looks like, a cardboard cube can be constructed to contain this volume of air, the length of each side being only a little over 28 cm. This cube would be able to contain, at s.t.p., 2 g hydrogen, H_2; 32 g oxygen, O_2; 71 g chlorine, Cl_2; 16 g methane, CH_4; 17 g ammonia, NH_3; etc. 6.0225×10^{23} molecules of each!

Atomicity

It may be wondered how the chemists of the nineteenth century decided that hydrogen was diatomic and that the formula of its molecule must be H_2. This was partly argued by using Avogadro's explanation of Gay-Lussac's Law of 1809 (Section 78). The law states that

When gases react, the volumes in which they do so are in a simple ratio to each other and to the volume(s) of the product(s), if also gaseous, provided that all volumes are measured at the same temperature and pressure.

The basic reaction, which was the starting point for determining the formulae of many other gases, was that between hydrogen and chlorine to form hydrogen chloride. It was determined by experiment that

1 vol. of hydrogen + *1 vol.* of chlorine give *2 vols* of hydrogen chloride

Applying Avogadro's hypothesis:

1 mol. of hydrogen + *1 mol.* of chlorine give *2 mols* of hydrogen chloride

Thus having fixed the ratio of the molecules involved we can write:

$$1H_x + 1Cl_y \longrightarrow 2H_aCl_b$$

where x and y represent the respective atomicities of hydrogen and chlorine and a and b are numbers to be determined.

Taking the simplest possible formula for hydrogen chloride, where a and b are each one, x and y both have to be 2 and the equation becomes

$$H_2 + Cl_2 \longrightarrow 2HCl$$

Later, there was also evidence from the **specific heat ratio**. The specific heat capacity (specific thermal capacity) of a gas is the heat required to raise the temperature of unit mass of the gas by 1 °C. This value can be determined either at constant volume (c_v) or at constant pressure (c_p). In the latter case, expansion of the gas takes place and more energy is needed to raise the temperature by 1 °C. c_p, therefore, always has a higher value than that for c_v. The ratio of c_p to c_v was found to have the highest value when the molecular state of the gas was the simplest, i.e. one atom per molecule. All monatomic gases, e.g. the noble gases (Section 24) have a specific heat ratio of about 1.66. Diatomic gases, including hydrogen, oxygen, nitrogen, carbon monoxide, and hydrogen chloride give a ratio of about 1.4. Triatomic gases, such as carbon dioxide, dinitrogen oxide, and steam have a specific heat ratio of about 1.3.

We now know that there are very good reasons why the atoms of elements such as hydrogen, chlorine, and oxygen form pairs and this is explained in Section 9.

Relationship between the relative density of a gas and its relative molecular mass

In order to determine the density of a gas relative to hydrogen, the masses of equal volumes of the two gases are compared under the same physical conditions. As these equal volumes must contain an equal number of molecules, this is equivalent to finding the mass of a molecule of the gas relative to the mass of a *molecule* of hydrogen. If the molecule of hydrogen is known to be diatomic the value for the relative density must be half the relative molecular mass with respect to an atom of hydrogen

The relative molecular mass of a gas is always twice the value for its relative density.

Examples are given in Table 2.

Table 2

Gas	Relative molecular mass	Relative density
Hydrogen, H_2	2	1
Chlorine, Cl_2	71	35.5
Hydrogen chloride, HCl	36.5	18.25
Oxygen, O_2	32	16
Steam, H_2O	18	9

It follows, also, that

the density of any diatomic gaseous element, relative to hydrogen, will be equal to its relative atomic mass.

Formula for a molecule of water (in the form of steam) and for a molecule of oxygen

Using apparatus, such as that described in Section 26, kept well above 100 °C, it can be shown that

$$2 \text{ vols hydrogen} + 1 \text{ vol. oxygen} \longrightarrow 2 \text{ vols steam}$$

Applying Avogadro's hypothesis:

$$2 \text{ mols hydrogen} + 1 \text{ mol. oxygen} \longrightarrow 2 \text{ mols steam}$$

Thus, having fixed the ratio of the molecules involved, we can write:

$$2H_2 + 1O_z \longrightarrow 2H_2O_c$$

where z is the atomicity of oxygen and c is a number to be determined.
The relative density of steam is 9 and its relative molecular mass is therefore 18. The tentative formula H_2O_c given above must correspond with a relative molecular mass of 18. So $2 + 16c$ must be equal to 18 and c therefore is 1. Thus the equation becomes:

$$2H_2 + O_2 \longrightarrow 2H_2O$$

THE MOLE CONCEPT

A hydrogen atom has a mass of approximately 1 unit compared with an atom of carbon (^{12}C) taken arbitrarily to have a mass of exactly 12. On this same scale of relative masses, a hydrogen molecule, H_2 has a mass of $1+1=2$, a water molecule H_2O, has a mass of $1+1+16=18$, and a sodium ion, Na^+, has a mass of 23. The removal of an electron from a sodium atom of mass 23 has no noticeable effect on the mass, since the electron itself has negligible mass. These relative masses have long been known although the actual masses of the incredibly small particles have only recently been calculated.

In practice, it is impossible for the chemist to handle single atoms, molecules or ions. Any sample that he studies will contain a vast number of particles. Consider samples of hydrogen atoms, hydrogen molecules, water molecules, and sodium ions each containing n particles (atoms, molecules or ions). Their masses will clearly be:

$$H\ 1n \qquad H_2\ 2n \qquad H_2O\ 18n \qquad Na^+\ 23n$$

on the same scale as before. It will be noted that the masses are in the same ratio $(1:2:18:23)$ as were the masses of the individual particles themselves.

In laboratory work, the usual unit of mass is the gram and so it is convenient to consider a sample of water, for example, in which the number of molecules is such that the sample has a mass of 18 grams. The value of n proves to be the Avogadro number, L. Because the relative masses stay the same regardless of the value of n, it follows that samples of hydrogen atoms, hydrogen molecules, water molecules, and sodium ions each containing the Avogadro number of particles would have masses:

$$H\ 1g \qquad H_2\ 2g \qquad H_2O\ 18g \qquad Na^+\ 23g$$

i.e. the numercial value of the mass in grams is equal to the relative atomic, molecular or ionic mass on the $^{12}C=12$ scale.

Formerly, these masses were given special names:

1 g of hydrogen atoms   = 1 gram-atom of hydrogen
2 g of hydrogen molecules = 1 gram-molecule of hydrogen
18 g of water molecules   = 1 gram-molecule of water
23 g of sodium ions   = 1 gram-ion of sodium

The separate terms given in the right-hand column above are no longer used; they all represent the same kind of quantity and are now all covered by the term **mole**. Thus

1 g of hydrogen atoms   = 1 mole of hydrogen atoms
2 g of hydrogen molecules = 1 mole of hydrogen molecules
18 g of water molecules   = 1 mole of water molecules
23 g of sodium ions   = 1 mole of sodium ions

In using the term 'mole' ambiguities must be carefully avoided. The expression 'one mole of hydrogen' is ambiguous in the light of the examples given. Does it mean one mole of hydrogen atoms (1 g) or one mole of hydrogen molecules (2 g)? Unless there were good reason to think otherwise, one would normally take 'one mole of hydrogen' to refer to the element in its normal state, i.e. molecules of H_2, but it is better, if there is any chance of ambiguity, to avoid this by giving the formula. Thus, in the example, 'one mole of hydrogen, H_2' would be preferable to 'one mole of hydrogen'.

4   Classification of Elements: Metals and Non-metals

Chemists have, during many centuries, made a rough distinction between metals and non-metals; but the division of the ninety-two natural elements into just two classes is not of much help as there are many elements that do not fit neatly into either class.

Typical metallic elements have certain characteristic properties, (see Table 3).

Table 3

Metals	Non-metals
Chemical	
Readily form positive ions (Section 8)	Negative ions or, more usually,
Compounds are therefore electrovalent (Section 8)	covalent compounds
Oxides are basic in character (Section 32)	Acidic or neutral
Chlorides are solids, stable, not easily hydrolysed (Section 20)	Covalent: most can be hydrolysed
Can replace hydrogen from non-oxidizing acids (Section 34)	Cannot do so
Do not form stable hydrides (Section 25)	Do form these
Act as reducing agents (Section 31)	Oxidizing agents
Physical	
Have a typical shiny appearance	Are not shiny
Conduct heat and electricity well	Do not conduct heat and electricity (graphite, Section 49, is an exception.)
Are malleable (easily beaten into shape)	Are not malleable
Are ductile (easily drawn into wire)	Are not ductile

To say that an element is metallic is equivalent to saying that it is electropositive in character (Section 7).

Looking at the chart of the Periodic Classification of the elements on pages 248–9 and, disregarding the noble gas group on the right-hand side, if a line is drawn across the chart from the top-left corner to the bottom-right corner, most of the metals will be found to be on the left of the line and the non-metals on the right. The most electropositive element (that is not radioactive), caesium, Cs, is in the bottom-left corner. The most electronegative element, fluorine, F, is in the extreme top-right corner. Near to where the line passes, will be found the indeterminate elements that are difficult to classify, e.g. hydrogen, H, aluminium, Al, germanium, Ge, arsenic, As.

Consider aluminium. Its physical characteristics are metallic but in aluminates it can appear as a negative ion. Many of its compounds are covalent. Its oxide is amphoteric (Section 32) neither markedly basic nor acidic. Its chloride is easily hydrolysed (Section 20).

Consider hydrogen. It has none of a metal's physical characteristics. Its ion is usually positive. Its oxide, H_2O, is neutral. Its chloride, HCl, is a stable covalent gas. Hydrogen is an important reducing agent.

Classification by electronegativity (Sections 6 and 7) is found to be much more helpful.

5   Classification of Elements: Periodic Classification

During the middle part of the nineteenth century, when many atomic weights (relative atomic masses) were determined, it was noted that when the atomic weight increases by a certain quantity the same chemical properties are observed again in the next element. In other words, a certain periodicity was noted in the atomic weight values.

In 1829, a German chemist, Döbereiner, had already pointed out that in certain 'triads' of elements, that possessed similar chemical properties, the atomic weights ascended in arithmetical progression, so that the value for the middle one was almost exactly between the other two. The examples he gave were:

Lithium	7	Calcium	40
Sodium	23	Strontium	87
Potassium	39	Barium	137

As the atomic weight increases by a certain number of units, the same chemical properties recur. This is the meaning of 'periodicity'.

In 1865, an Englishman named Newlands, carried the idea much further and, by comparison with a musical scale, published his 'Law of Octaves'. By writing out the atomic weights in ascending order and starting another row at every *eighth* element, those of similar chemical properties were to be found grouped together in vertical columns; just as, in a musical scale a related note (the octave) comes at every eighth note.

In 1871, the Russian chemist Mendeléeff, published his well-known Periodic Classification of the elements. It was not unlike the table used today but now chemists prefer to list the elements by **atomic number** (Section 2) rather than by atomic weight. Mendeléeff had many problems in compiling his table: Where should hydrogen be placed? How could the 'transition elements' be-fitted in? One or two elements, listed by atomic weight, seemed to be in the wrong places. Some atomic weights had to be corrected. Most of the difficulties disappeared when, in the twentieth century, classification was made by atomic number. In the table used today there is, at the right-hand end, one more group – the noble gases (Section 24). These were not discovered until 1895 and, if Newlands had known about them, he would have had to start a new row (series) at every *ninth* element and his exact analogy with the musical scale would not have been valid.

As there are ninety-two natural elements on the periodic chart and the relative atomic mass of the last one, uranium, is 238 it follows that the average difference between the masses of one element and the next is about $2\frac{1}{2}$. Several elements had not been isolated in Mendeléeff's time and, as there was a gap of about ten units of mass between zinc and arsenic, he found that he had to leave two spaces for numbers 31 and 32. By considering the properties of elements above and below, he was able to make brilliant predictions concerning the properties, both chemical and physical, of these missing elements. When gallium and germanium were eventually isolated, in 1875 and 1886 respectively, their properties were found to correspond almost exactly with those that Mendeléeff had forecast. The position of hydrogen is still a problem. With a valence of 1 one would expect it to be either with Group I elements (very electropositive) or Group VII elements (very electronegative) (Sections 6 and 7), but its properties do not agree with either of these.

Looking at a modern periodic chart (pages 248–9) it will be seen that there are eight main groups. Complications arise when the transition series start at element 21 but some explanation of this is given in Section 70. Students should memorize the positions of the first twenty elements, i.e. as far as calcium.

Group I is known as the *Alkali Metal group*, because each of the elements (disregarding hydrogen) reacts readily with water to form an alkali (Section 22).

Group II is known as the *Alkaline Earth group*. These metals have some properties in common with those of Group I.

Group III The only common element here is aluminium. Boron is not well known but some of its compounds such as borax and boric acid are in everyday use.

Group IV shows in an interesting manner the change from non-metal, carbon, to metal, lead, as the relative atomic mass increases from element to element.

Group V has nitrogen and phosphorus (non-metals), arsenic and antimony (not definitely one or the other), and bismuth (metal).

Group VI starts with oxygen and sulphur. Although these two elements appear to have little similarity, their compounds have much in common.

Group VII is known as the *Halogen group* because each of these elements combines directly with a metal to give a salt (Section 23).

Group VIII The *Noble Gas group*. A group of relatively inert gases which are sometimes classed as Group 0 (Section 24).

Between element 57 (lanthanum) and element 72 (hafnium) a series of fourteen uncommon elements, the 'lanthanons', has been omitted. Elements 93 to 106 have been artificially created and are radioactive.

The very important difference that marks the modern Periodic Classification from that produced by Mendeléeff is that the twentieth-century classification is based on **atomic structure** (Section 2). The first series contains only two elements because the first energy level of electrons can hold just two – hydrogen has one, helium has two – before that level is filled. The second energy level can accommodate a maximum of eight electrons, so the second series runs thus:

Li 2,1   Be 2,2   B 2,3   C 2,4   N 2,5   O 2,6   F 2,7   Ne 2,8

Two energy levels of electrons are now filled. The third energy level can contain 18 electrons and starts by filling up to eight again thus:

Na 2,8,1   Mg 2,8,2   Al 2,8,3   Si 2,8,4   P 2,8,5   S 2,8,6   Cl 2,8,7   Ar 2,8,8

Potassium, K has an electron configuration 2,8,8,1 and calcium, Ca, 2,8,8,2. After this the next ten electrons (after element 20) enter the third energy level again giving the first transition series (Section 70).

In the nineteenth century the **valence** (or valency) of an element was defined simply as 'the number of hydrogen atoms that one atom of the element can combine with or replace'. When the Periodic Classification was adopted, it was immediately seen that there is a very important connection between group and valence. For group I elements the valence is one, for Group II the valence is two, Group III has valence three, and Group IV valence four. After this the valence goes up *and* down. Group V elements show valences of five and three, Group VI six and two, and Group VII seven (not common) and one. Group VIII elements have no valence.

With the elucidation of the structure of atoms, valence was shown to be the result of electron behaviour and a more meaningful understanding of valence became possible (Sections 8 and 9). All elements that have one electron in the outer energy level of their atoms will have similar properties (Alkali Metals, Section 22). All elements that have two electrons only in the outer energy level of their atoms will resemble each other (Alkaline Earth Metals), and so on. This is the fundamental explanation of the observed 'periodicity' in the properties of the elements.

6 Classification of Elements: Activity Series

Another very helpful way of classifying the commoner elements is by the study of their behaviour and chemical reactivity. Table 4 below lists twenty-nine elements that are dealt with in this book. The list is known as an **activity series** or sometimes as the **electrochemical series.**

Table 4   The activity (or electrochemical) series

Caesium	Cs	*Most electropositive*
Potassium	K	
Sodium	Na	
Lithium	Li	
Barium	Ba	
Strontium	Sr	
Calcium	Ca	
Magnesium	Mg	
Aluminium	Al	
Zinc	Zn	
Iron	Fe	
Tin	Sn	
Lead	Pb	
Hydrogen	H	
Copper	Cu	
Mercury	Hg	
Silver	Ag	
Platinum	Pt	
Gold	Au	
Silicon	Si	
Phosphorus	P	
Iodine	I	
Carbon	C	
Sulphur	S	
Bromine	Br	
Nitrogen	N	
Chlorine	Cl	
Oxygen	O	
Fluorine	F	*Most electronegative*

The dotted line in Table 4 indicates where the elements that are commonly regarded as non-metals begin. The list is so useful that students might do well to copy it onto a strip of cardboard for use as a book-mark. It will often be referred to in the course of this book.

The series is also a table of relative electronegativity (Section 7).

One drawback to the use of this list is that, unlike the Periodic Classification, chemists do not entirely agree about the relative positions of some of the elements in the series. There are several ways of fixing the position of any one element in the table but, unfortunately results are not always consistent.

Methods used for determining the relative order of the elements

(1) REPLACEMENT: A metal higher in the list will replace one that is lower from an aqueous solution of a salt of the latter metal. Thus zinc will replace copper from a solution of copper(II) sulphate but will not replace magnesium from a solution of magnesium sulphate. Again, copper will replace mercury from a solution of mercury(II) chloride but will not replace tin from a solution of tin(II) chloride. It will be seen later (Section 31) that these reactions are examples of oxidation/reduction. When one metal replaces another, the further they are apart on this list, the greater is the energy change, i.e. more heat is evolved.

(2) REACTION WITH WATER: The metals at the top of the list react vigorously, even explosively, with cold water. The vigour of the reaction falls off as the hydroxide of the metal concerned becomes less soluble, the metal becoming coated with a protective layer. Calcium hydroxide is not very soluble in water, so the reaction is less dangerous. Magnesium hydroxide is almost insoluble and so the reaction with cold water is very slight. The next five metals (aluminium to lead) may react when heated in steam, but with progressively less vigour. In all cases where there is a reaction, hydrogen gas is liberated. Metals that come below hydrogen in the list are unable to replace it and these have no action with either water or steam. Copper is the cheapest of these and hence is much used for making water pipes, condenser tubes, etc. (Section 72).

(3) REACTION WITH DILUTE NON-OXIDIZING ACIDS (hydrochloric or sulphuric): The alkali metals react dangerously with these acids, but hydrogen is replaced by all metals above hydrogen but with less vigour as the table is descended. Notable exceptions are aluminium (protective layer of oxide (Section 42)) and lead (insolubility of salt formed (Section 44)).

(4) EASE OF FORMATION AND STABILITY OF THE OXIDE: Those metals furthest from oxygen react most readily with it and the oxides formed are correspondingly stable, i.e. they are not decomposed by ordinary heating and are not easily reduced to the metallic state. Potassium even forms a superoxide, KO_2. By contrast, copper heated in air forms only a superficial layer of copper(II) oxide and this oxide is readily reduced to copper. Mercury(II) oxide is very easily decomposed just by heating. Platinum and gold do not combine with oxygen.

(5) ELECTRODE POTENTIALS: If a metal rod is placed in an aqueous solution of one of its salts (of standard concentration and standard temperature – usually 25 °C) a potential difference is set up between the rod and the solution. This is known as the **electrode potential.** By combining this half-cell with a standard hydrogen electrode (value fixed arbitrarily as 0.00 V) its value can be measured. If we consider the potential of the metal with respect to the solution, the more active the metal the greater the number of ions (Section 8) which go into solution and the more negative the metal rod becomes. This explains why, in Table 5, the most electropositive metal has the highest negative value. The figures obtained follow the activity series (or perhaps in this case it would be better termed the electrochemical series). High figures at the top reduce progressively to zero at hydrogen and then the figures increase again but with the opposite sign. It is possible to construct gas electrodes for hydrogen, chlorine, etc. Table 5 shows some of the values obtained.

Table 5

Element	Electrode potential/V
Lithium	−3.02
Calcium	−2.76
Aluminium	−1.66
Zinc	−0.76
Iron	−0.44
Lead	−0.13
Hydrogen	0.00
Copper	+0.34
Silver	+0.80
Chlorine	+1.36
Fluorine	+2.85

When Volta constructed the first voltaic cell (about 1800) he chose the metals zinc and copper (or silver) and dipped them in dilute sulphuric acid. Using copper the e.m.f. he obtained would be the difference between the two values given above, i.e. -0.76 and $+0.34 = 1.10$ V. Later, Leclanché chose the elements zinc and carbon (graphite) for his cell and obtained an e.m.f. of about 1.5 V. This was to be expected because carbon is lower in the electrochemical series than copper. The further apart the chosen elements are in the series, the greater the e.m.f. obtainable. Theoretically therefore a cell using lithium and gold would be better still but practical difficulties, including expense, rule out many elements from general use.

Occurrence in nature

The most reactive elements never occur in nature in the free state. The first metal in the list to be found so is copper and this in only a relatively small quantity. Silver and gold (often called the 'noble metals' because of their relative inactivity) are found free in nature.

(See also Section 7.)

7   Electronegativity

Electronegativity has been defined, by Pauling, as '*the power of an atom in a molecule to attract electrons to itself*'. Much of what was written in Section 6 depends upon this power. Nevertheless it is difficult to measure electronegativity in the laboratory. Pauling, Mulliken, and others, acting independently, have found methods of assessing relative electronegativities and the results do produce rough agreement, the order of the elements being similar to the list given in Section 6.

The atoms of electropositive (metallic) elements tend to lose electrons, whereas the atoms of electronegative elements tend to gain them. (The term 'electropositivity' is not used, only relative electronegativity.) From Sections 5 and 6, fluorine should have the highest value and caesium the lowest. Table 6 gives some approximate values, expressed on an arbitrary scale, for the relative electronegativities of twenty common elements:

Table 6

Element	Relative electronegativity
Caesium	0.7
Potassium	0.8
Sodium, Barium	0.9
Calcium	1.0
Magnesium	1.2
Aluminium	1.5
Zinc, Iron	1.65
Tin	1.7
Hydrogen, Phosphorus	2.1
Iodine	2.4
Carbon, Sulphur	2.5
Bromine	2.8
Nitrogen, Chlorine	3.0
Oxygen	3.5
Fluorine	4.0

When a compound is formed from *two* elements only, it is known as a **binary** compound. The further the elements are apart on this scale of relative electronegativity, the stronger the bond between the two atoms will be and also the more ionic its character (Sections 8 and 9). Thus the strongest possible bond would be between caesium and fluorine when caesium fluoride, CsF, is formed. (The ending '-ide' is always used when naming a binary compound – the more electropositive element being put first.) On the other hand, we should not expect the bonds holding the atoms together in hydrogen iodide, HI, or a chlorine oxide, CO_2, to be very strong.

The concept of electronegativity is of great importance and the terms 'electropositive' and 'electronegative' are much used in this book. Students should realize that there is nothing frightening about this concept and the electronegativity series is much the same as the activity series listed in Section 6.

Relationship between Periodic Classification and electronegativity

Although these methods of classifying the elements have quite separate histories, and were evolved by quite different means, there is an important relationship between them. If one looks at the electronegativity series, one sees that the alkali metals appear in the order: caesium, potassium, sodium, lithium – which is the sequence in which they appear in Group I of the Periodic Table if read upwards from the bottom. Again, in Group II, the order of the elements in the electronegativity scale is: barium, strontium, calcium, magnesium – which, once again, is the order of Group II metals in the Periodic Table read upwards. One might be tempted to think, therefore, that one could write out a complete electronegativity series by listing the elements from each group of the Periodic Table in turn, reading from bottom to top. The situation is not, however, as simple as this because, apart from some overlapping, the elements of Groups III, IV, and V tend to share electrons rather than to transfer them (Section 9). The transition series of elements (Section 70) also offer further complications. Looking at the right-hand side of the Periodic Table, it can be seen that the halogen elements there appear in the reverse order to that in the electronegativity scale: iodine, bromine, chlorine, fluorine. Likewise, in Group VI, sulphur comes above oxygen.

Knowing now the structure of the atom, it is easy to explain these facts. Electrons in shells that are completely filled play no part in chemical reaction. It is the electrons of unfilled shells that are responsible for chemical activity. All the elements of Group I, the alkali metals, have one electron only in the outermost shells of their atoms. This gives them all a valence of 1 because it is this outermost electron, in each case, that is detached. This explains why all these metals are electropositive. It can readily be shown that it is easier to detach an electron from a caesium atom than from a lithium atom. Consider the structure of each, remembering that the atomic volume is, of course, three-dimensional and not possible to show on a printed page:

Caesium Z (atomic number) $= 55$ Relative atomic mass $= 133$
    Nucleus   *Electron configuration* $(55-)$
$\begin{cases} 55 \text{ protons} + \\ 78 \text{ neutrons} \end{cases}$ $2 : 8 : 18 : 18 : 8 : 1$

Lithium $Z = 3$ Relative atomic mass 7
    Nucleus   *Electron configuration* $(3-)$
$\begin{cases} 3 \text{ protons} + \\ 4 \text{ neutrons} \end{cases}$ $2 : 1$

The valence electron of the lithium atom is relatively close to the nucleus, with its three positive charges, and is only being repelled by the two electrons in the first shell. The valence electron of the caesium atom, on the other hand, is relatively far away from the positive nucleus and is being repelled by the negative charges of fifty-four electrons in the inner shells. Thus it is much easier to detach the valence electron from the caesium atom – which is therefore the more electropositive. Exactly similar reasoning can be used to show why potassium is more electropositive than sodium and so on. In Group II of the Periodic Table, each metal has two valence electrons and it can be shown, in the same way, that it should be easier to detach the two electrons from the barium atom than from the calcium atom.

The Group VII (halogen) elements have seven electrons each in the outermost shells of their atoms. These elements are electronegative because their atoms tend to attract an electron each to complete the outermost shells. Again, one can readily show that fluorine will be much more electronegative than iodine simply by considering the atomic structure:

Fluorine $Z = 9$ Relative atomic mass $= 19$

Nucleus	*Electron configuration* $(9-)$
9 protons $+$ 10 neutrons	$2 : 7$

Iodine $Z = 53$ Relative atomic mass $= 127$

Nucleus	*Electron configuration* $(53-)$
53 protons $+$ 74 neutrons	$2 : 8 : 18 : 18 : 7$

An electron approaching the fluorine atom will be attracted by the relatively close positive nucleus and repelled only by the charges of 9 electrons. An electron approaching the iodine atom, by contrast, will be relatively distant from the positive nucleus and repelled by 53 negative charges. Clearly then, it will be more difficult for the iodine atom to accept another electron than for the fluorine atom to do so. Similar argument will show why chlorine is more electronegative than bromine. In the same way, in Group VI, where the atoms each have six electrons in their outermost shells and tend to accept two more electrons, it is possible to argue that oxygen must be more electronegative than sulphur.

If one follows the elements along one *period* of the Periodic Classification, for example from sodium to chlorine, it will be seen that, as one passes from left to right, each succeeding element become more electronegative. If we consider a *group*, reading downwards, say lithium to caesium or fluorine to iodine, each succeeding element becomes more electropositive.

Thinking along these lines will help the student to remember the position of common elements in both systems of classification.

8   Linkage of Atoms: Electrovalent (ionic)

A linkage between atoms of two elements to form a compound usually involves two electrons forming a pair. If an atom, or group of atoms, has an unpaired electron it is likely to be reactive, combining readily with a second atom or group. If we consider the simple case of an atom A combining with another atom B to form a compound AB, then there are three important ways in which the atoms can be linked. These are:

(1) An electron from A can be *transferred* from atom A to atom B. A and B then exist not as atoms but as *ions*. The bond is **electrovalent.**
(2) Atom A and atom B each contribute one electron and the electron pair is *shared* between them, thus making a **covalent** bond (Section 9).
(3) Atom A *or* atom B contributes *both* electrons which are then shared between the two atoms, making a special covalent bond sometimes known as a dative ('giving') covalent bond, but more often termed a **co-ordinate** linkage (Section 10)

In this section the first of these is considered.

Electrovalent (or *ionic*) linkage

This bond is most likely when an electropositive element from Groups I or II combines with an electronegative element from Groups VI or VII of the Periodic Classification. The electropositive elements can act as electron donors and the electronegative elements as electron acceptors.

When a sodium atom combines with a chlorine atom to form the binary compound sodium chloride, the one valence electron of the sodium atom is transferred to the chlorine atom which has seven electrons in its outer shell. (The first two shells of electrons, already completed, need not be considered.) When this transfer has been made, the sodium atom $(Z=11)$ is left with 11 protons but only 10 electrons. It therefore has a surplus positive charge and is known as an **ion.** (An *ion* may be defined as an *atom, or group of atoms, that has gained or lost one or more electrons.*) Likewise the chlorine atom $(Z=17)$ receives an extra electron and becomes an ion with a negative charge, thus

$$\text{Na}\cdot + \ :\!\overset{\displaystyle ..}{\underset{\displaystyle ..}{\text{Cl}}}\!: \ \longrightarrow \ \text{Na}^+ \quad :\!\overset{\displaystyle ..}{\underset{\displaystyle ..}{\text{Cl}}}\!:^-$$

$$\text{2,8,1} \qquad \text{2,8,7} \qquad\qquad \text{2,8} \qquad \text{2,8,8}$$

The resulting compound is very stable and the ions are held together by strong electrostatic attraction. It is important to realize that, in sodium chloride, there are *no separate molecules* of NaCl but the whole is built into a crystal (cubic shaped in this case) which is a **giant molecule of ions** arranged in a rigid **lattice.** In this lattice each sodium ion is surrounded by six chloride ions and each chloride ion by six sodium ions. It is thus impossible to allot any sodium ion to any particular chloride ion. The formula NaCl represents the simplest (empirical) formula which expresses no more than that for every sodium ion there must be chloride ion (Figure 2).

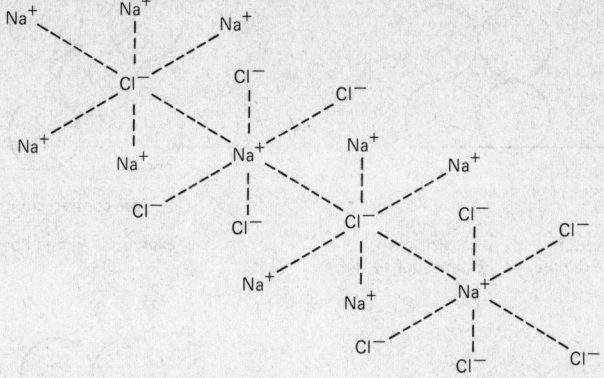

Figure 2   Lattice structure of sodium chloride, NaCl

Similarly, when magnesium chloride is formed, each of the two valence electrons of the magnesium atom $(Z=12)$ is transferred to a chlorine atom, giving a crystal lattice with the empirical formula $MgCl_2$ $(Mg^{2+}2Cl^-)$:

$$Mg \; + \; \begin{matrix} :\ddot{C}l: \\ 2,8,7 \\ :\ddot{C}l: \\ 2,8,7 \end{matrix} \longrightarrow Mg^{2+} \; + \; \begin{matrix} :\ddot{C}l:^- \\ 2,8,8 \\ :\ddot{C}l:^- \\ 2,8,8 \end{matrix}$$

Mg
2,8,2

Mg²⁺
2,8

The oxides of the elements of Groups I and II are ionic in character. When calcium oxide is formed by burning calcium $(Z=20)$ in oxygen $(Z=8)$ two electrons are transferred from the calcium atom to the oxygen atom, which has only six electrons (two unpaired) thus:

$$Ca \; + \; :\ddot{O}: \longrightarrow Ca^{2+} \; :\ddot{O}:^{2-}$$

2,8,8,2   2,6   2,8,8   2,8

One should not speak of electrons 'wanting' or 'trying' to do this or that because, as far as we know, electrons have no feelings but are governed by natural laws. The law in this case is the filling of shells to form the inactive noble-gas structure.

Electrovalent (ionic) compounds formed by the transfer of one or more electrons from atom to atom, can usually be recognized by their possession of several important properties:

(1) They *are solids having rigid crystalline lattice structures.* Often the structure will be one of four fairly simple types: (a) simple cubic, (b) face-centred cubic, (c) body-centred cubic or (d) hexagonal (see Figure 3).

(2) As the electrostatic attractions hold the ions in position very firmly, *melting points and boiling points are high.* It will be remembered that sodium chloride was, to the alchemist, an 'element' that could not be changed – the 'principle of permanence' (Section 1).

(3) *They are readily soluble in water but not, usually, in organic solvents.* To

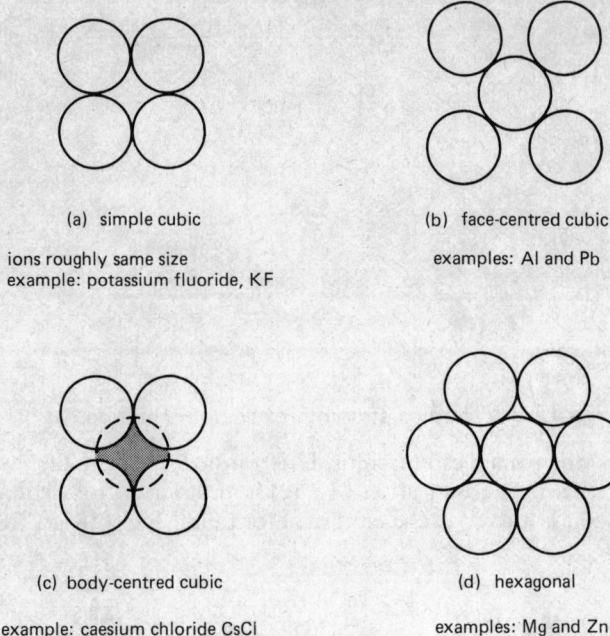

(a) simple cubic

ions roughly same size
example: potassium fluoride, KF

(b) face-centred cubic

examples: Al and Pb

(c) body-centred cubic

example: caesium chloride CsCl

(d) hexagonal

examples: Mg and Zn

Figure 3　Crystalline lattice structures

make a solution, the electrostatic forces between the ions have to be weakened, giving the ions freedom of movement. This can be done only by those liquids that have high dielectric constants. Of all the common liquids water has the highest value and most organic liquids have low values. A few other liquids have high enough values to act as ionizing solvents (e.g. liquid ammonia, hydrogen cyanide, HCN) but these are not common. The solvent, however, does *not* form the ions. The ions are present in the crystals and an ionizing solvent simply weakens the forces that hold the ions together.

(4) *Ionic crystals, in the solid state, cannot conduct electricity* because the ions are held so rigidly in position. When dissolved in an ionizing solvent or fused (melted by heating), the ions are free to move and, by their movement, conduct readily.

Common ions that involve groups of atoms are the ammonium, NH_4^+ (Section 10), the nitrate, NO_3^-, and the sulphate SO_4^{2-}. Whenever an atom or a group has a charge sign we know at once that it is an ion and not an atom or a molecule. *A positive ion is always smaller than the atom from which it has been formed, whereas a negative ion is larger than its corresponding atom.* To take a simple case: sodium has an electron configuration 2, 8, 1. When its one valence electron is transferred elsewhere, it is left with only two shells of electrons and a further contraction takes place because the remaining electrons are pulled more strongly towards the nucleus by the surplus positive charge.

9 Linkage of atoms: Covalent

The atoms of elements from Groups III, IV, and V of the Periodic Classification are more likely to form covalent than ionic bonds for a very simple reason. When an atom has accepted an electron to become an ion with one negative charge, it will be more difficult for it to accept a second electron because of the repulsive effect of the electron already gained. However, this does frequently happen. For a third or fourth electron to be added is much more difficult and it is unlikely that there would ever be an ion of charge 4− or 4+. This is one of the reasons why the **Stock notation** has been adopted, which gives the valence of an element in Roman numerals. Tin, for example, can have valencies of 2 or 4. Divalent tin (formerly called 'stannous') could be written Sn(II) or as Sn^{2+} because it is ionic in character. Tetravalent tin (formerly called 'stannic') *must* be written Sn(IV) because the tin atom could not lose as many as four electrons to form an ion.

Most of the compounds formed by aluminium (Group III), carbon and silicon (Group IV), nitrogen and phosphorus (Group V) are covalent – where two atoms combine by *sharing* an electron pair to which each atom has contributed one electron. A very simple case is the gas methane, CH_4. The carbon atom has four valence electrons ($Z=6$) and each electron pairs with an electron provided by one of the four hydrogen atoms; which results in the formation of four covalent bonds directed to the four corners of a tetrahedron. Except in combination with *very* electropositive elements, as in the hydride of sodium Na^+H^-, the hydrides of other elements are usually covalent, e.g. HCl, H_2O, H_2S. The formation of the last one, hydrogen sulphide, can be shown thus:

$$H\cdot \ + \ H\cdot \ + \ :\overset{\cdot\cdot}{\underset{\cdot\cdot}{S}}: \ \longrightarrow \ \overset{\textstyle H}{\underset{\textstyle H}{:\overset{\cdot\cdot}{\underset{\cdot\cdot}{S}}:}}$$

Covalent compounds, formed by the *sharing* of electron pairs between atoms, also have characteristic properties which distinguish them from electrovalent (ionic) compounds. The most important of these are:

(1) *They are usually liquid or gaseous and consist of discrete molecules.* Students often wonder why many common gaseous elements are written as having two atoms to the molecule (diatomic), such as H_2, Cl_2, O_2. Each of these atoms has 1 (or in the case of oxygen 2) unpaired electron and, this being reactive, two atoms join thus:

$$H\cdot \ + \ H\cdot \ \longrightarrow \ H:H \quad (H_2)$$

$$:\overset{\cdot\cdot}{\underset{\cdot\cdot}{Cl}}\cdot \ + \ \cdot\overset{\cdot\cdot}{\underset{\cdot\cdot}{Cl}}: \ \longrightarrow \ :\overset{\cdot\cdot}{\underset{\cdot\cdot}{Cl}}:\overset{\cdot\cdot}{\underset{\cdot\cdot}{Cl}}: \quad (Cl_2)$$

$$:\overset{\cdot\cdot}{\underset{\cdot}{O}}: \ + \ :\overset{\cdot\cdot}{\underset{\cdot}{O}}: \ \longrightarrow \ :\overset{\cdot\cdot}{O}:\overset{\cdot\cdot}{\underset{\cdot\cdot}{O}}: \quad (O_2)$$

(See also Section 3)

In the case of the noble gases, the atoms have no unpaired electrons and therefore exist as single atoms (monatomic). Most common gaseous elements are diatomic or monatomic and only rarely triatomic, e.g. ozone, O_3.

(2) Because covalent molecules have no, or only a slight, dipole they are easily separated from one another, i.e. *they have low melting and boiling points.*

(3) *They are not usually soluble in water, but will dissolve readily in organic solvents* such as carbon disulphide, CS_2, tetrachloromethane (carbon tetrachloride), CCl_4, ethanol, ethoxyethane (diethyl ether), etc.

(4) Because there are no ions, covalent compounds *cannot conduct electricity.*

(5) *Molecules* of covalent compounds *have definite shapes* which, in many cases, can be predicted. Electron pairs are negatively charged and have a repelling effect on one another. Therefore, if confined in a given volume (the atomic volume) they will tend to take up positions as far from one another as possible. For one atom of an element combining covalently with two atoms of another element, the shape could well be linear; with three atoms of another, the shape could be triangular; with four atoms, tetrahedral; with five atoms, a trigonal bipyramid; and with six atoms of another element, octahedral. Examples of all these shapes are known:

Carbon dioxide, CO_2 is *linear.*

$$:\ddot{O}=C=\ddot{O}:$$

each line representing an electron pair. Boron trichloride, BCl_3, is *triangular.*

Methane, CH_4 is *tetrahedral.*

Phosphorus pentabromide, PBr_5, is *trigonal bipyramidal.*

Sulphur hexafluoride, SF_6 is *octahedral.*

Shapes are not always predictable and the commonest covalent compound, water, H_2O, is not a linear compound. The two hydrogen atoms attached to the oxygen atom are at rather more than a right angle:

$$\begin{array}{c} H \\ \diagdown \; 105° \\ :\underset{\cdot\cdot}{O}\!\!-\!\!H \end{array}$$

It is not possible to classify all compounds as rigidly electrovalent (ionic) or covalent. For example, monochloromethane, CH_3Cl, is mainly covalent in character but also has a slight ionic nature. Hydrogen atoms in organic compounds (Sections 53 and 54) may be looked upon as neutral atoms because their presence makes the molecule neither electropositive nor electronegative. Methane, CH_4, (above) is a typically covalent compound. It has no dipole. If, however, an atom of the very electronegative element, chlorine, is substituted for one of the hydrogen atoms to give monochloromethane, CH_3Cl, the molecule immediately gains a dipole and hence a slightly ionic character.

10   Linkage of Atoms: Co-ordinate

Important examples of the co-ordinate linkage (sometimes called the 'dative covalent bond') are provided by water and by ammonia. The bond is similar to a covalent one but the electron pair is contributed by only one of the two atoms. The linkage is not strong and, in many cases, can be broken by energy supplied by gentle heating. As it involves the *transfer* of an electron pair from one atom to the other the product will be **polar,** i.e. the molecules will have positive and negative 'ends'—negative where the atom has accepted electrons and positive where the other atom has donated them. Indeed, it has been called a 'co-ionic' bond.

Water

The water molecule has two electron pairs which are not attached to any other atom and one, or both, of these *could* be used to form a co-ordinate link with another atom that happens to be an electron pair short of the normal complement:

$$\begin{array}{c} H \\ \diagdown \\ :\underset{\cdot\cdot}{O}\!\!-\!\!H \end{array}$$

These electron pairs are known as **lone pairs.**

The oxonium ion (also sometimes called the 'hydroxonium' or 'hydronium' ion) A hydrogen ion has no electrons; the one electron that the atom had was lost when it became an ion, so forming a proton H^+. It is now known that a proton does not exist in aqueous solution because it immediately attaches itself to a water molecule forming the oxonium ion, H_3O^+:

$$H^+ + H_2O \longrightarrow H_3O^+ \text{ (oxonium ion)}$$

$$H^+ + \ :\overset{\displaystyle H}{\underset{\displaystyle\cdot\cdot}{\text{O}}}:H \longrightarrow \left[\ :\overset{\displaystyle H}{\underset{\displaystyle H}{\text{O}}}:H\ \right]^+$$

It is unlikely that another proton would attach itself to the other spare electron pair because the second proton would be repelled by the positive charge already there.

The lone pairs of electrons in water molecules also come into play in the formation of *hydrates* (Section 29).

The ammonium ion

When a molecule of ammonia is formed from atoms of nitrogen and hydrogen, the nitrogen atom still has one lone pair of electrons. The nitrogen atom has five valence electrons and only three of these unite with electrons from three hydrogen atoms to form three covalent bonds. Ammonia itself is, of course, a covalent compound:

$$\cdot\overset{\cdot}{\text{N}}: \ + \ 3H\cdot \ \longrightarrow \ H:\overset{\displaystyle H}{\underset{\displaystyle H}{\text{N}}}: \ \text{(the lone pair)}$$

The great importance of this lone pair lies in the fact that, under the right conditions, an ammonia molecule can accept a proton to form an ammonium ion, NH_4^+.
It is therefore a base (Section 35).

$$H^+ + NH_3 \longrightarrow \left[\ H:\overset{\displaystyle H}{\underset{\displaystyle H}{\text{N}}}:H\ \right]^+$$

Ammonia molecules can also take the place of water molecules in a hydrate to form an *ammine* (Section 29).

Most salt hydrates lose their water when gently heated and ammonium chloride on heating readily dissociates (Section 63):

$$NH_4Cl(s) \longrightarrow NH_3(g) + HCl(g)$$

11   Linkage of Atoms: Hydrogen Bond

As a hydrogen atom has only one electron one would not expect it to be capable of forming a linkage between two other atoms, but chemists are now sure that it *is* possible for one hydrogen atom to act as a *bridge* between two atoms of a very electronegative element. Hydrogen bonds, therefore, are only encountered when dealing with elements at the bottom end of the activity series, e.g. oxygen and fluorine. Hydrogen bonding is used to explain the crystal structure of ice and the relatively high boiling point of water. This special type of bond is represented by dotted lines and it is suggested that water molecules are linked with one another in this way, the bonds breaking progressively as the temperature rises from below 0 °C to 100 °C. One can represent such bonding thus:

$$
\begin{array}{c}
\quad\quad\quad H \\
\quad\quad\quad | \\
O-H\cdots O-H\cdots \\
| \\
H \\
\\
O-H\cdots O-H\cdots \\
| \quad\quad\quad | \\
H \quad\quad\quad H
\end{array}
$$

It is well known that hydrogen fluoride molecules, HF, tend to link together (associate) to form chains and this is also explained by the existence of hydrogen bonding:

$$
\cdots H_{\diagdown F}\cdots H^{\diagup F}\cdots H_{\diagdown F}\cdots H^{\diagup F}\cdots
$$

The fact that hydrofluoric acid is a much weaker acid than hydrochloric acid is attributed to the tendency of hydrogen fluoride molecules to associate rather than to dissociate (Sections 23 and 33).

12   Linkage of Atoms: Metallic

The structure of solid metals is definitely crystalline although, except when metal is freshly cast, it is difficult for the eye to see. The arrangements of the atoms are similar to those in crystals of common compounds:

Face-centred cubic includes aluminium, lead, copper, and silver.

Body-centred cubic includes the alkali metals and barium.

Hexagonal structure includes magnesium, zinc and mercury (in solid form).

Most metals are mechanically strong and have relatively high melting points. They are also good conductors of electricity. Such properties do *not* accord either with electrovalent or covalent bonding because not only must the linkages be as strong as these but free movement of electrons from atom to atom *must* be possible in order to allow the free passage of electricity through the solid. There must be an almost instaneous exchange of valence electrons from atom to atom when a current flows. Explanations of this phenomenon have been given, but, at this stage, it is sufficient to realize that metallic bonds must differ from the other linkages described. (Sections 8–11).

13   States of Matter. Kinetic Theory

STATES OF MATTER

Of the ninety-two natural elements, eleven are gases, two are liquids, and seventy-nine are solids, under ordinary conditions. Solid elements will melt when heated sufficiently (in an inert atmosphere if necessary) and at still higher temperatures they can be vaporized. Liquids, similarly, can be readily vaporised by heating and by cooling they can be solidified (mercury becomes solid when immersed in liquid oxygen). Gases can be liquefied by compression and/or cooling or by more specialized means. In other words, all the elements can under the right conditions, be obtained in any of the three states. There is no real distinction between *gas* and *vapour* except that the gaseous form of an element that is commonly known in the liquid or solid state, is called a vapour. Thus we speak of water vapour because water is normally a liquid and of iodine vapour because iodine is normally a solid. Vapours and gases obey the same natural laws.

KINETIC THEORY

Gases

The word 'kinetic' comes from the Greek 'kinesis' which means 'movement'. The kinetic theory assumes that the molecules of a gas are always in random movement and the velocity of movement depend upon the temperature – the hotter the gas, the faster the movement. At absolute zero $(-273°C$ $(0\,K))$ all movement ceases and all gases become liquid, then solid, before that temperature is reached. Collisions between molecules can be ignored, unless the gas is very compressed because the molecules of a gas (or vapour) are extremely small compared with the space in which they are moving and the assumption is made that all collisions are elastic – no energy being lost. When enclosed in a container, the impact of the molecules on the walls of the container produces the gas pressure. Boyle's and Charles' Gas Laws can be derived from the kinetic theory of gases. Boyle's Law tells us that doubling the pressure on a gas, at constant temperature, will halve its volume, but even very high pressures have little effect on liquids or solids. The gas form takes up a very much larger space than the liquid from which it is derived. $1\,cm^3$ of water when vaporized at ordinary pressures produces nearly $1700\,cm^3$ of steam. *Ideal gases* should obey Boyle's and Charles' Laws even at high pressures and low temperatures, but no gas is truly ideal. Hydrogen and helium come nearest to ideal behaviour.

Two other gas laws can be deduced from the kinetic theory: Dalton's Law of Partial Pressures and Graham's Law of Diffusion.

Dalton's Law states that when a mixture of gases, which do not chemically react with one another (such as air), occupies an enclosed space each gas will exert the pressure that it would exert if it occupied the space by itself;

the total pressure, therefore, is the sum of the partial pressures of the gases present in the mixture. A simple example of the application of this law, taken from elementary chemistry, concerns the collection of gases over water at a certain temperature and atmospheric pressure. The gas will be damp, i.e. holding as much water vapour as it is able to at the temperature concerned. The atmospheric pressure is balanced by the pressure of the gas plus the small pressure due to the water vapour. Therefore, to obtain the correct pressure of the gas, the saturation vapour pressure of water at the particular temperature has to be *subtracted* from the atmospheric pressure.

Diffusion is the tendency that a gas or vapour has to spread out in all directions and this, of course, is due to the random movement of the molecules. Graham, in 1833, discovered that less dense gases travel faster through a porous partition than more dense ones. He stated a useful quantitative law that provides one method of determining the relative molecular mass of a gas.

Graham's Law states that gases diffuse at rates that are inversely proportional to the square roots of their respective densities.

For gases A and B this can be expressed in mathematical form:

$$\frac{\text{Rate of gas A}}{\text{Rate of gas B}} = \sqrt{\frac{\text{Density of B}}{\text{Density of A}}}$$

Scientists find it easier to measure not the rates but the times taken by two different gases to diffuse under identical conditions.

For equal volumes of two gases to diffuse under the same conditions, the times taken are directly proportional to the square roots of their respective densities

because a gas that diffuses at twice the rate of another will take half the time. Graham's Law, expressed in terms of times taken, then becomes:

$$\frac{\text{Time taken by A}}{\text{Time taken by B}} = \frac{\sqrt{d_A}}{\sqrt{d_B}}$$

As the density of a gas, relative to hydrogen, is always half the relative molecular mass of the gas (Section 3), Graham's Law provided a method of finding the relative molecular mass of a given gas. For example, if it is found that, under identical conditions, chlorine takes $1\frac{1}{2}$ times longer to diffuse through a porous plug than does an equal volume of oxygen (relative density known to be 16), substitution in the above equation will give a value for the relative density of chlorine of 36. Therefore, the relative molecular mass, from this experiment, has a value of 72.

Diffusion provided one of the first methods for separating the isotopes of a gaseous element (Section 2).

Liquids

At ordinary temperatures and pressures all pure liquids, apart from the elements bromine and mercury, are compounds, most of which are covalent. They are virtually incompressible but offer no resistance to change of shape. The molecules have freedom of movement, but not the full freedom of movement that the molecules of a gas have, because the liquid in any container is bounded by a surface. The molecules at the surface are under tension because, although they are affected by the attractive forces of molecules below them and on the surface, there is not the same attraction above them. In an enclosed vessel, molecules are continually escaping from the surface into the vapour above and molecules are also returning from the vapour to the liquid. At constant temperature an equilibrium is set up in which molecules escape in numbers equal to those returning. Thus the liquid has a vapour pressure. When the temperature is raised, the molecules move faster and the vapour pressure increases. At boiling point, in an open vessel, the vapour pressure is equal to atmospheric pressure and molecules escape perfectly freely. Below the boiling point, in an open vessel, the liquid loses molecules into the atmosphere, i.e. it evaporates at a rate determined by the temperature. For a liquid of relatively high boiling point, such as lubricating oil, evaporation at room temperature is slow, but with ethoxyethane (ordinary ether) evaporation is very rapid because its boiling point is only 34.5 °C.

When a solid ionic compound is turned into liquid form by heating strongly, i.e. melted (or 'fused'), the ionic lattice structure is still loosely preserved though the ions have much greater freedom of movement.

Solids

Under ordinary conditions solids include metals, a few non-metallic elements, electrovalent compounds, and many organic compounds whose relative molecular masses are high (usually above 100). These last are large molecules with little freedom of movement. Diamond (Section 49) has a very rigid structure of covalent bonding. Solids are generally incompressible and offer resistance to change of shape, though such resistance varies widely from solid to solid. The movement of the constituent particles – whether atoms, ions, or molecules – is limited to vibratory motion. The vapour pressure of a solid is usually very low and there is no measurable evaporation under ordinary conditions. There are exceptions, as ice can evaporate slowly in the open air (without first melting) and naphthalene, $C_{10}H_8$, can decrease, steadily and obviously, in bulk by evaporation when left in the open air.

In the past a number of solids have been described as **amorphous**, i.e. having no shape or non-crystalline, but the number of these has been much reduced as greater knowledge concerning the solid state has been gained. Some, like glass (Section 61) have proved to be supercooled liquids, i.e. liquids that have cooled under conditions that did not allow crystallization to take place and others have been shown to be imperfectly crystalline, such as 'amorphous' carbon (Section 49).

14   Mixtures

Mixtures differ from compounds in many ways. When a particular compound is formed from its elements there is always an energy change, heat being evolved in most cases. The product always has the same composition by mass. For example, when sodium chloride is formed by combination of its elements, much heat is evolved and the proportion is always one ion of sodium to one ion of chlorine, i.e. $23:35.5$ by mass giving 58.5 NaCl. The product formed by chemical union has very different properties from those of the elements that have been used to form it. Consider how different the properties of sodium chloride (common salt) are from those of sodium or chlorine! To recover the sodium and the chlorine from the compound formed needs much energy to melt and electrolyse the sodium chloride (Section 39).

By contrast, when a mixture is made:

(1) *There is no observable energy change.*
(2) *The components can be mingled in any proportions.*
(3) *The properties of the mixture are the sum of the properties of the substances in it.* Each component retains its own characteristic properties.
(4) *The ingredients can be separated by physical means.*

Separation of mixtures

Any number of ingredients can be used in producing mixtures and some natural ones, such as the soil or milk, can be very complicated. Consideration is only given here to all possible binary (two-component) mixtures and the simple means by which they can be separated.

SOLID/SOLID: It is usually possible to find a solvent that will dissolve one and not the other. The mixture is filtered. The insoluble one can be washed on the filter paper and the filtrate evaporated to recover the other.
In special cases sublimation can be used (Section 19).
In certain cases both solids can be dissolved in one solvent and then separated by chromatography.

SOLID/LIQUID: Filtration, or use of a centrifuge.

SOLID/GAS: Such a mixture occurs when small particles of solid are suspended in the air, flue gases, etc. The solid can be removed by special filters or by electrostatic precipitation.

LIQUID/LIQUID:
Non-miscible: These liquids do not mix to form one homogeneous phase. ('Homogeneous' means 'the same all over'. The opposite term is 'heterogeneous'.) If two such liquids are shaken together vigorously they may, for a short while, appear to be homogeneous (forming what is known as an emulsion) but they will soon form two separate layers. One can be drawn away from the other with the aid of a separating funnel. Example: carbon disulphide/water.

Miscible: If the two liquids have boiling points far enough apart, ordinary distillation may separate them reasonably well. If the boiling points are fairly close fractional distillation might be used. In some cases chromatographic methods would serve.

LIQUID/GAS: If a gas is dissolved in a liquid, heating to boil the liquid for about ten minutes will usually completely expel the gas. If fine particles of liquid are dispersed over gas (as in fog) special filters or electrostatic precipitation can be used.

GAS/GAS: It is usually possible to find a solid or a liquid (pure or solution) that will absorb one gas but not the other. Often the first can be recovered by simple means afterwards. Examples:
Water vapour can be absorbed by quicklime or concentrated sulphuric acid. Carbon dioxide is best absorbed by potassium hydroxide solution. Oxygen can be absorbed by an alkaline pyrogallate solution (Section 30). Water may be used to absorb any very soluble gas, e.g. ammonia, hydrogen chloride, sulphur dioxide.

When elements or compounds dissolve in water, without chemical action, to form solutions, these also are mixtures (Section 28).

15   Purification of Common Substances

Solids

Most solids can be crystallized from a suitable solvent. The faster the crystallization takes place, the smaller and purer are the crystals. If large crystals are formed by slow evaporation of the solvent impurities are likely to occur in the crystal structure. To obtain purer crystals, a hot saturated solution is quickly cooled. The crystals precipitate and, after filtering and washing, can be dissolved again and recrystallized from the pure solvent. A few solids can be purified by sublimation (Section 19).

Liquids

After separating (non-miscible liquids) (Section 14) or distilling (miscible ones) the purity can be checked by determining at least two physical properties, such as density and freezing point (or boiling point). To show that a liquid is pure water, for example, it must be shown that its density is 1 g cm^{-3} and its b.p. exactly $100\,°\text{C}$ at $760\,\text{mm}$ pressure (or f.p. $0\,°\text{C}$). Water will turn the colour of anhydrous copper(II) sulphate from white to blue, but this is only a test for the *presence* of water. Any substance that has a small quantity of water with it will turn anhydrous copper(II) sulphate blue. If ethanol turns this compound blue it simply shows that the ethanol is not water-free. It does *not* prove that the liquid is pure water.

Liquids are often dried by being allowed to stand over a compound that will absorb water, e.g. quicklime (calcium oxide) or anhydrous calcium chloride.

Gases

A gas is usually purified by passing it through various absorbing agents in separate containers, each one chosen to abstract a particular impurity which is likely to be present. Finally, the gas is dried, usually by passing it through concentrated sulphuric acid or anhydrous calcium chloride. The alkaline gas, ammonia, must be dried through calcium oxide. When dry, a gas must be collected by displacement of air, if possible, *not over water*. Gases are seldom collected over mercury, except on a very small scale, because the liquid is *very* expensive and very dense (bee-hive shelves would float on it!).

16   Formulae

The valence of an element is determined by the number of electrons (valence electrons) in the outermost shell of its atom (Sections 8–10). If the atom readily loses only one electron (e.g. an alkali metal, Section 22) the element is monovalent. Similarly, an element will be monovalent if its atom readily gains an electron (e.g. a halogen, Section 23). (A transition element is a special case and can have more than one valence, Section 70.) If the atom of an element can readily lose two electrons (e.g. a Group II metal) or readily gain two electrons (e.g. a Group VI element such as sulphur), that element will be divalent.

Common monovalent metals are sodium, Na, potassium, K, and silver, Ag. *One* atom of each of these metals can, under the right conditions, replace *one* atom of hydrogen (also monovalent). The group of atoms NH_4, the ammonium radical (Section 10), can also act as a monovalent unit.

Common divalent metals are magnesium, Mg, calcium, Ca, barium, Ba, lead, Pb, and zinc, Zn. *One* atom of each of these metals can, under the right conditions, replace *two* atoms of hydrogen.

One well-known metal that has a valence of III (trivalent) is aluminium, Al. *One* atom of this element can replace *three* atoms of hydrogen.

Iron, a transition element (Section 71), has two common valencies: iron(II) (ferrous) and iron(III) (ferric).

Copper, also a transition element (Section 72) has two common valencies: copper(I) (cuprous) and copper(II) (cupric).

A **hydroxide** can be regarded as a compound in which *one* of the two hydrogen atoms in a molecule of water, H.OH, has been replaced by a metal. Thus,

sodium hydroxide is   Na̶HOH   NaOH

calcium hydroxide is   Ca $\begin{matrix} H̶OH \\ H̶OH \end{matrix}$ $Ca(OH)_2$

aluminium hydroxide is   Al $\begin{matrix} H̶OH \\ H̶OH \\ H̶OH \end{matrix}$ $Al(OH)_3$

An **oxide** can be regarded as a compound in which *both* hydrogen atoms of the water molecule have been replaced by a metal. Thus,

  sodium oxide is Na_2O
  calcium oxide is CaO

and   aluminium oxide is Al_2O_3 because *two* atoms of aluminium can replace *six* atoms of hydrogen

$$Al_2 \begin{matrix} H̶OH̶ \\ H̶OH̶ \\ H̶OH̶ \end{matrix}$$

Carbonates are derived from carbonic acid, H_2CO_3, which is a dibasic acid (Section 33). If *one* hydrogen atom of the carbonic acid molecule is replaced

Table 7 Formulae of common compounds

	Element		Hydroxide	Oxide	Hydrogen-carbonate	Normal carbonate	Chloride	Nitrate	Hydrogen sulphate	Normal sulphate	Normal sulphide
Monovalent	Sodium	Na	NaOH	Na_2O	$NaHCO_3$	Na_2CO_3	NaCl	$NaNO_3$	$NaHSO_4$	Na_2SO_4	Na_2S
	Potassium	K	KOH	K_2O	$KHCO_3$	K_2CO_3	KCl	KNO_3	$KHSO_4$	K_2SO_4	K_2S
	Silver	Ag	—	Ag_2O	—	Ag_2CO_3	AgCl	$AgNO_3$	—	Ag_2SO_4	Ag_2S
	Ammonium	NH_4	NH_4OH	—	NH_4HCO_3	$(NH_4)_2CO_3$	NH_4Cl	NH_4NO_3	NH_4HSO_4	$(NH_4)_2SO_4$	$(NH_4)_2S$
Divalent	Magnesium	Mg	$Mg(OH)_2$	MgO	$Mg(HCO_3)_2$	$MgCO_3$	$MgCl_2$	$Mg(NO_3)_2$	—	$MgSO_4$	MgS
	Calcium	Ca	$Ca(OH)_2$	CaO	$Ca(HCO_3)_2$	$CaCO_3$	$CaCl_2$	$Ca(NO_3)_2$	—	$CaSO_4$	CaS
	Barium	Ba	$Ba(OH)_2$	BaO	—	$BaCO_3$	$BaCl_2$	$Ba(NO_3)_2$	—	$BaSO_4$	BaS
	Lead	Pb	$Pb(OH)_2$	PbO	—	$PbCO_3$	$PbCl_2$	$Pb(NO_3)_2$	—	$PbSO_4$	PbS
	Zinc	Zn	$Zn(OH)_2$	ZnO	—	$ZnCO_3$	$ZnCl_2$	$Zn(NO_3)_2$	—	$ZnSO_4$	ZnS
Trivalent	Aluminium	Al	$Al(OH)_3$	Al_2O_3	—	—	$AlCl_3$	$Al(NO_3)_3$	—	$Al_2(SO_4)_3$	Al_2S_3
Transition elements	Iron(II)	Fe	$Fe(OH)_2$	—	$Fe(HCO_3)_2$	$FeCO_3$	$FeCl_2$	—	—	$FeSO_4$	FeS
	Iron(III)	Fe	$Fe(OH)_3$	Fe_2O_3	—	—	$FeCl_3$	$Fe(NO_3)_3$	—	$Fe_2(SO_4)_3$	Fe_2S_3
	Copper(I)	Cu	—	Cu_2O	—	—	CuCl	—	—	—	Cu_2S
	Copper(II)	Cu	$Cu(OH)_2$	CuO	—	$CuCO_3$	$CuCl_2$	$Cu(NO_3)_2$	—	$CuSO_4$	CuS

'—' indicates that the compound does not exist or is uncommon by a metal, a 'hydrogencarbonate' (bicarbonate) is formed:

$$\text{sodium hydrogencarbonate} \quad NaHCO_3$$
$$\text{calcium hydrogencarbonate} \quad Ca(HCO_3)_2$$

If *both* the hydrogen atoms are replaced a normal carbonate is formed:

$$\text{sodium carbonate} \quad Na_2CO_3$$
$$\text{calcium carbonate} \quad CaCO_3$$

Sulphates are derived from sulphuric acid, H_2SO_4 (also dibasic); sodium hydrogensulphate, $NaHSO_4$ (bisulphate); normal sodium sulphate, Na_2SO_4.

Chlorides are derived from hydrochloric acid, HCl.

Nitrates are derived from nitric acid, HNO_3.

Normal sulphides are derived from hydrogen sulphide, H_2S. Hydrogensulphides, e.g. NaHS, are also known.

17   Equations: Molecular

It was believed at one time that, in any chemical reaction, molecules were disrupted into atoms which then rearranged themselves to form new molecules. A molecular equation showed, on the left-hand side, the molecules present before the reaction and, on the right-hand side, the new molecules present after the reaction. All atoms must be accounted for because chemists know that atoms cannot be created or destroyed by any chemical means. In other words, the equation must be balanced. Since the recognition of different types of atomic linkage, we know that the complete picture of chemical reaction given above only applies to reactions that are mainly between covalent substances. In electrovalent compounds, separate molecules do not exist and the reactions are ionic in character (Section 18).

In writing molecular equations the large figure in front of a formula indicates the number of molecules, e.g. $2H_2O$ represents two molecules of water that in reaction can give four atoms of hydrogen and two atoms of oxygen. It is not necessary to put the figure 1 in front when only one molecule is involved. It is often helpful to indicate whether the substance is in solid (s), liquid (l) or gas (g) form.

Molecular equations will always be required for gas reactions because all common gases exist as separate molecules. Likewise, as all common organic compounds are mainly covalent, molecular equations will be found throughout Sections 49–59. Here are a few varied examples of equations requiring molecular formulae:

Action of heat on copper(II) carbonate giving copper(II) oxide and carbon dioxide.

$$CuCO_3(s) \longrightarrow CuO(s) + CO_2(g)$$

Action of steam on red-hot iron to give the black oxide (Sections 48 and 71) and hydrogen.

$$3Fe(s) + 4H_2O(g) \longrightarrow Fe_3O_4(s) + 4H_2(g)$$

Reaction between sulphur dioxide and damp hydrogen sulphide gases to give water and sulphur.

$$SO_2(g) + 2H_2S(g) \longrightarrow 2H_2O(l) + 3S(s)$$

Complete combustion of ethene to give carbon dioxide and steam.

$$C_2H_4(g) + 3O_2(g) \longrightarrow 2CO_2(g) + 2H_2O(g)$$

Formation of ethanol and carbon dioxide by fermentation of a glucose solution.

$$C_6H_{12}O_6(aq) \longrightarrow 2C_2H_5OH(aq) + 2CO_2(g)$$

'(aq)' following a formula indicates that the substance is in aqueous solution (from the Latin 'aqua' meaning 'water').

Chemists now know that reaction can only take place when molecules come near enough to one another with the requisite energy (activation energy) (Section 47) and it is extremely unlikely that more than three molecules could react together at the same moment. Molecular equations that show larger numbers, e.g. three of one substance acting with eight of another, must either be incorrect or indicate the final result after two or three simpler reactions have taken place in rapid sequence.

18   Equations: Ionic. Ionic Aggregation

Reactions that take place between electrovalent compounds can be recorded in a much simpler manner. Electrovalent compounds have no molecules, only a lattice of ions that are rigidly fixed in the solid form. When dissolved in water, the ions are free to move and react.

Consider the reaction between solutions of sodium chloride and silver nitrate. This would formerly have been written in molecular form:

$$NaCl(aq) + AgNO_3(aq) \longrightarrow AgCl(s) + NaNO_3(aq)$$

If we consider the ions present, there are $Na^+(aq)$, $Cl^-(aq)$, $Ag^+(aq)$, and $NO_3^-(aq)$. When the solutions are brought together all that happens is that the silver ions unite with the chloride ions to form insoluble silver chloride which leaves the system as a precipitate. The sodium ions and nitrate ions play no part and are sometimes referred to as 'spectator' ions. All we need write then to show the change that has taken place is:

$$Ag^+(aq) + Cl^-(aq) \longrightarrow AgCl(s)$$

Such precipitation reactions used to be known as 'double decomposition'. This term is not now acceptable because it perpetuates the idea of the molecules of two compounds decomposing (Section 17). The term 'metathesis' (from the Latin meaning 'interchange') has also been used, but this is not much better. It is true that, if solid sodium nitrate is recovered from the solution as well as the precipitated silver chloride, then the chloride and the nitrate groups have changed partners. The best term to describe what has happened is **ion aggregation.** Such reactions are very useful for the preparation of insoluble compounds (Section 37).

Further examples of ion aggregation
The test for sulphate ion with barium chloride solution. Consider magnesium sulphate, $MgSO_4$ with barium chloride, $BaCl_2$. The ions present are Mg^{2+}, SO_4^{2-}, Ba^{2+}, and $2Cl^-$.

All that happens is

$$Ba^{2+}(aq) + SO_4^{2-}(aq) \longrightarrow BaSO_4(s)$$

The magnesium and chloride are spectator ions.

Consider a slightly more difficult example. A solution of ammonia, NH_4OH, added to iron(III) chloride solution, precipitates insoluble iron(III) hydroxide. The molecular equation would be:

$$FeCl_3 + 3NH_4OH \longrightarrow Fe(OH)_3 + 3NH_4Cl$$

The ions present are $Fe^{3+}(aq)$, $3Cl^-(aq)$, $3NH_4^+(aq)$, and $3OH^-(aq)$.
The simple ionic equation is therefore:

$$Fe^{3+}(aq) + 3OH^-(aq) \longrightarrow Fe(OH)_3(s)$$

19   Decomposition of Compounds: Pyrolysis

'Pyro' comes from the Greek word for 'fire' and **pyrolysis** means 'the splitting up of a compound by means of heat', i.e. **thermal decomposition.** The hottest flame from a laboratory burner may heat an object to 800 °C or more and this is sufficient to break up many compounds. ('Red-heat' is about 400–450 °C and 'white-heat' is about 650 °C). The compound does not as a rule break down into elements. Only a few, such as thermally unstable oxides (e.g. mercury and silver) or iodides, being binary compounds formed from elements fairly close to one another in the activity series, break down readily to give the constituent elements. During pyrolysis, one or more gases are often evolved and the process is irreversible (Section 48).

Compounds formed from elements at the extreme ends of the activity series, such as the fluorides, oxides, and chlorides of Groups I and II, are very stable thermally and are not affected by temperatures up to 1000 °C. Nearly all compounds of the alkali metals are thermally stable – sulphates, hydroxides, and even carbonates are unaffected by ordinary heating, though the hydroxide and carbonate of lithium (the least electropositive of the group) are decomposed. Group I nitrates form nitrites on heating (Section 64) but most other nitrates give the oxide, nitrogen dioxide and oxygen. The crystals of lead nitrate, which do not contain water of hydration, break up violently when heated – a phenomenon known as **decrepitation** ('crackling'). Sulphates of the less electropositive metals can be decomposed, for example, copper(II) sulphate gives copper(II) oxide and sulphur trioxide. Hydroxides and carbonates of the less electropositive metals decompose readily, giving oxide and steam in the first case and oxide and carbon dioxide in the second.

Gentle heating is sufficient to release water from hydrates (Section 29) and to cause **dissociation** and **sublimation** of particular compounds where co-ordinate linkages are involved.

Substances that dissociate form simpler molecules that recombine on cooling.

If a solid dissociates completely into a gas or a vapour, the molecules forming simpler molecules, their recombination reforms the solid. Substances that do this, e.g. iodine (Section 23) and ammonium chloride (Section 63) can be purified by this means which is, in effect, a dry distillation. The process is called **sublimation** and the solid reformed on cooling is called a **sublimate.**

Very strong heating can dissociate electrovalent compounds to a slight extent into the constituent ions and this enables some metals to be recognized by 'flame tests'. The colours associated with various elements, chiefly those of Groups I and II, are:

Yellow (bright and persistent)	Sodium	Red	Calcium
		Crimson	Lithium or strontium
Pink/blue (lilac)	Potassium	Green	Barium or copper

20 Decomposition of Compounds: Hydrolysis

Hydrolysis is the decomposition of a compound through the agency of water. When the hydrolysis is complete, one part of the compound combines with the hydrogen from the water and the other part with the hydroxyl group.

Salts of metals in Groups I and II which have been formed from *strong* acids do *not* hydrolyse, but many examples can be found among the mainly covalent compounds formed from elements in Groups III, IV, and V. As examples, aluminium chloride and phosphorus(III) chloride are readily hydrolysed by cold water, producing fumes of hydrochloric acid as the hydrogen chloride reacts with water vapour in the air:

$$AlCl_3(s) + 3H.OH(l) \longrightarrow Al(OH)_3(s) + 3HCl(g)$$
$$PCl_3(l) + 3H.OH(l) \longrightarrow P(OH)_3(aq) + 3HCl(g)$$

In the second case, because $P(OH)_3$ is an acid, it is usually written with the hydrogen first, H_3PO_3 (phosphorous acid) (Section 66).

Another familiar example, from organic chemistry, is the formation of ethyne (acetylene) gas by the action of water on calcium carbide (Section 75).

$$CaC_2(s) + 2H.OH(l) \longrightarrow Ca(OH)_2(s) + C_2H_2(g)$$

Hydrolysis may be brought about in various ways using (a) cold water, as in the examples above; (b) hot water; (c) water heated under pressure (thus raising the boiling point); (d) steam; (e) superheated steam – each method being more potent than the ones preceding it. The later ones are common in organic chemistry where hydrolysis is often slow and more difficult to bring about. In some cases, dilute acid or alkali can produce hydrolysis more effectively than water itself.

It is important to distinguish between hydrolysis and hydration. In the latter case there is *no* decomposition but simply the *addition* of water molecules to an ion to form a hydrate (Section 29); but, in many cases, hydration may well precede hydrolysis.

Many salts, not including those formed from strong acids and strong alkalis, are hydrolysed to a slight extent when dissolved in, or particularly when heated with water. Water, because it is a compound of hydrogen and hydroxyl, can behave either as a very weak acid or as a very weak alkali. It cannot compete with strong acids or alkalis and, in consequence, sodium chloride, for example, is not hydrolysed. Copper(II) sulphate, on the other hand, *is* affected to a slight extent by water and a heated solution will become cloudy because of the slight formation of insoluble copper(II) hydroxide and sulphuric acid. The acid is ionized and the solution has an acid reaction with an indicator. Sodium ethanoate, on the other hand, is slightly hydrolysed into ethanoic acid, which is weak and provides hardly any hydrogen ions, and sodium hydroxide, which is completely ionized, giving an excess of hydroxide ions. Sodium ethanoate solution, therefore, gives an alkaline reaction with indicators and so also does sodium carbonate solution, for the same reasons.

21   Decomposition of Compounds: Electrolysis

Electrolysis is the breaking up of a compound by passing an electric current through it.

After Volta had discovered in 1800 how to produce an electric current by means of a voltaic cell, a battery of such cells was soon used to decompose water into its elements (Section 26). Later, Davy used electrolysis to isolate, for the first time, the metals potassium, sodium (in 1807), and magnesium. Now it has become one of the most important processes used by the chemist, both in the laboratory and in industry.

Any ionic compound can be electrolysed either in the form of a melted solid or as a solution. This is known as the **electrolyte.** Into this are placed two rods or plates to act as the **electrodes.** The cell that contains a solution to be electrolysed is always tall and narrow in order to minimize resistance between the plates. Direct current *must* be used, either from a battery or from the mains supply with a transformer and a rectifier. Alternating current will achieve nothing because the change caused by the current passing in one direction is immediately and exactly reversed by the current flow in the opposite direction. Even with direct current, nothing will happen unless the difference in potential exceeds, for the particular electrolyte, its decomposition potential, which may be over 3 V. Above that figure, the higher the voltage the greater the current and the more rapid the decomposition. For ordinary laboratory experiments, 6 to 12 V is satisfactory. The positive electrode is known as the **anode** and the negative electrode as the **cathode.** Electrolysis can be a very complicated process and, even with the same electrolyte, different products can be obtained under varied conditions. The products obtained may depend on the nature of the electrodes, the concentration of the electrolyte, the current density, and the temperature.

When a direct current is passed through an electrolyte, any positive ions (metals or hydrogen) drift towards the cathode and are known as **cations.** Negative ions (**anions**), such as Cl^-, OH^-, SO_4^{2-}, drift towards the anode. When the cations reach the cathode surface they take up electrons to become neutral atoms. This is a process of reduction (Section 31). If hydrogen is discharged, two atoms at once form a molecule; if a metal, it is likely to plate onto the cathode. If there are two or more different metals present there will be a preferential discharge the order of which can be predicted from the electrode potential series (Section 6). The less electropositive will be deposited before the more electropositive, e.g. copper before zinc if ions of both these metals are present. The anions, when they reach the surface of the anode, give up electrons to form atoms (an oxidation process) and then, usually, molecules. If chlorine is to be discharged, a metal cannot be used as the anode because chlorine attacks all metals, so graphite anodes have to be employed. If there are two or more different anions present in the solution, again there will be preferential discharge, the less electronegative first, e.g. iodine before chlorine.

Faraday introduced the now familiar terms set bold in the first part of this section and in 1834 stated two important quantitative laws.

1. *The chemical power of a current of electricity is in direct proportion to the absolute quantity of electricity which passes.* Expressing this in modern terms: *The mass of an ion liberated in electrolysis is directly proportional to the quantity of electricity passed.*

The quantity is measured in coulombs, one coulomb (C) being a current of one ampere (A) passed for one second. Thus if a current of 2 A is passed for five minutes (i.e. 600 C) through an electrolytic cell, the decomposition that takes place would be exactly the same as if a current of 10 A had been passed for one minute (also 600 C).

Secondly, Faraday found that

2. *It takes 96 500 C of electricity to liberate the gram-equivalent of any element. This quantity is known as the* **Faraday constant** *and is given the symbol* F.

Nowadays, chemists prefer to work in moles (Section 3) rather than in equivalents, but the value of the equivalent is easily obtained by dividing the relative atomic mass of an element (expressed in grams) by the valence that it has in the compound under consideration. The masses in grams of common elements liberated by one Faraday of electricity are shown in Table 8.

Therefore one Faraday liberates one mole of atoms of a monovalent element, one half-mole of atoms of a divalent element, and so on. Because one Faraday of electricity is able to liberate one mole of atoms of a monovalent element during electrolysis, it is regarded as being *one mole of electrons*. A gram atom (the relative atomic mass expressed in grams) contains the Avogadro Number

Table 8

Element (valence state)	Relative atomic mass	Gram-equivalent/g
Sodium(I)	23	23
Magnesium(II)	24	12
Aluminium(III)	27	9
Silver(I)	108	108
Copper(II)	63.5	31.75
Chlorine(I)	35.5	35.5
Oxygen(II)	16	8

of atoms (Section 3) and each, for a monovalent element, can accept or lose one electron. The atoms of a divalent element can accept or lose two electrons, so the gram atom of a divalent element will need two Faradays for liberation at the electrode.

Non-metallic solids, with the exception of graphite (Section 49), cannot conduct an electric current. A liquid (fused solid or solution) can only conduct if ions are present and, when it conducts, decomposition is inevitable. Gases can only conduct if ionized in some way.

Examples of the usefulness of electrolysis can be found in the following sections:

Extraction of metals:
  From fused solid – sodium (39), magnesium (41), aluminium (42)
  From solution – zinc (45), copper (72).
  Purification of metals: lead (44), copper (72)
  Plating of metals: with tin (43), zinc (45), and other metals (73).
  Protective oxide coating: aluminium (42)
  Large-scale preparation of elements and compounds: fluorine (23); sodium hydroxide; hydrogen, chlorine, hypochlorites, chlorates (38); hydrogen peroxide (32).

22   Families of Elements: The Alkali Metals

Sections 22–24 are concerned with three groups of the Periodic Classification: Group I, very electropositive elements; Group VII, very electronegative elements; and Group O (or VIII), inert (inactive) gases. The second member of each group is considered to be the most typical and if the properties of sodium are known (for Group I) and those of chlorine (for Group VII) the behaviour of the other members can be predicted. The first member of a group often shows some unexpected properties.

Group I has six members but the last three – rubidium, caesium, and francium – are rare and therefore only the first three – lithium, sodium, and potassium will be considered. As was shown in Section 7, the elements of this group become more electropositive as the atomic number increases.

Physical properties

The metals are alike in many ways:
Relatively soft metals that can be cut with a knife.
Less dense than water – float on it when reacting.
Relative densities at 20 °C; Li, 0.53; Na, 0.97; K, 0.86.
Melting points relatively low: Li, 186 °C; Na, 98 °C; K, 63 °C.

Chemical properties

The great similarity here is due to the fact that each has only one electron in the outermost shell of its atom. They all form electrovalent (ionic) compounds, therefore, and even the oxides and hydroxides are ionic. It is difficult to decompose these compounds by heating, though the nitrates do form nitrites, losing some oxygen. All the metals are monovalent. The elements can be obtained by electrolytic methods though, for various reasons, potassium is now obtained by a thermal method carried out in vacuo. The bright surface exposed when a metal is cut rapidly tarnishes because of rapid superficial oxidation. Alkali metals are usually kept under naphtha (a paraffin oil) to shield them from air and dampness, but in most oils lithium will float. All three metals give characteristic flame tests (Section 19). That the vigour of chemical reactivity falls off from potassium to lithium is convincingly shown by the action with oxygen and with water.

(1) **WITH OXYGEN**: When heated in oxygen potassium forms a very unusual compound, KO_2, known as a 'superoxide'. Sodium forms a peroxide, Na_2O_2 (Section 32). Only lithium forms the oxide one would expect from a monovalent element, Li_2O.

(2) **WITH WATER**: A small piece of potassium acts almost explosively with cold water to form K^+ and OH^- ions, hydrogen gas being liberated. The temperature is high enough to ignite the hydrogen which burns with a flame coloured lilac by the presence of potassium ions, even when the metal is darting round on the surface of the water.

A fragment of sodium behaves similarly but the hydrogen does not usually ignite (bright yellow flame) unless the water is warm or the fragment is placed on a filter paper to stop the fragment darting round on the surface.

Lithium reacts with water much more gently. The hydrogen, if ignited, burns with a crimson flame and the metal does not melt as the others do.

Each alkali metal dissolves in mercury to form an **amalgam.** Sodium amalgam is often used in the laboratory because it is safer than sodium alone.

Although the salts of potassium are much more costly than those of sodium – because there is not a cheap source of the element such as sodium has in common salt – they are usually preferred as reagents in the laboratory; the nitrate, iodide, bromide, chlorate, dichromate, manganate(VII) are very common. Sodium compounds are, as a rule, more soluble in water than those of potassium and many take up water from the air and become damp (i.e. they are **hygroscopic**) or even take up enough to form a solution (i.e. they are **deliquescent**), e.g. sodium hydroxide. Sodium salts, therefore, are not easy to keep pure. Sodium nitrate, for example, would be useless as a constituent of gunpowder. Pure sodium chloride is not hygroscopic but impure salt often contains deliquescent magnesium chloride as an impurity and this does become damp in moist air. All the common compounds of sodium and potassium are soluble in water.

Lithium, being the least electropositive of the alkali metals, shows some properties that are similar to those of the metals of Group II. Its carbonate and hydroxide are almost insoluble in water and no hydrogencarbonate (bicarbonate) can be obtained in solid form (compare magnesium and calcium). The carbonate and hydroxide each decomposes on **pyrolysis** to form the oxide (again similar to magnesium and calcium). Its halides (except the fluoride) and nitrate are, like those of magnesium and calcium, extremely deliquescent.

23   Families of Elements: The Halogens

There are five members of Group VII of the Periodic Classification but the uncommon and radioactive one – astatine – will not be considered. Each of the others – fluorine, chlorine, bromine, iodine – will combine directly with a metal to give a salt which is a binary compound known as a **halide**. The name **halogen** is derived from the Greek 'hals' and means 'source of salt'. Chlorine is the typical member of the group.

Each member of this group has atoms that have seven electrons in the outermost shells. Each is monovalent and readily accepts an electron from an alkali metal atom. All the members are electronegative but the electronegativity diminishes as the atomic number increases (Section 7). Fluorine is the most electronegative of all elements but iodine has a few electropositive properties, e.g. it forms an ionic compound with chlorine, I^+Cl^-.

Physical properties

(at ordinary temperatures and pressures)

Fluorine: Pale yellow/green *gas*. Rather denser than air (19:14.4).
Chlorine: Yellow/green *gas*. Much denser than air (35.5:14.4).
Bromine: Dark red/brown *liquid*. Three times denser than water.
Iodine: Black crystalline *solid*. Density $4.93 \, \mathrm{g\,cm^{-3}}$.

As the atomic number increases the elements become denser and darker in colour. All the elements are extremely poisonous, bromine and liquefied chlorine being especially dangerous. Fluorine gas, chlorine gas, and bromine vapour have diatomic molecules (Section 3) F_2, Cl_2, Br_2. Iodine vapour (which has a deep-violet colour), I_2, dissociates into atoms when heated; above 1700 °C it is monatomic:

$$I_2 \rightleftharpoons 2I.$$

Chemical properties

REACTION WITH HYDROGEN: Each combines with hydrogen but much less readily as the atomic number of the element increases. Each forms a covalent gas: HF, hydrogen fluoride; HCl, hydrogen chloride; HBr, hydrogen bromide; HI, hydrogen iodide. For example, with chlorine:

$$H_2 + Cl_2 \longrightarrow 2HCl$$

Hydrogen and fluorine combine explosively, even in the dark. Hydrogen and chlorine combine explosively, only in sunlight or when the mixture is ignited. In sunlight (**photochemical** reaction) the combination proceeds by a **chain**

reaction. A molecule of chlorine absorbs a quantum of energy which dissociates it into atoms:

$$Cl_2 \longrightarrow 2Cl$$

The chlorine atoms then initiate the chain reaction that proceeds thus:

$$Cl + H_2 \longrightarrow HCl + H$$
$$H + Cl_2 \longrightarrow HCl + Cl$$

Hydrogen and bromine vapour combine only when heated in the presence of a catalyst (e.g. platinum).

Hydrogen and iodine combine under the same conditions but the reaction is incomplete. Both the bromide and iodide formations are reversible (Section 48).

HALOGEN HYDRACIDS HF, HCl, HBr, HI: The hydrogen halides do not have acid properties until they combine with water, when hydrogen ions are at once formed. They are called hydracids because there are other acids that contain oxygen as well, e.g. hypochlorous acid, $HClO$ (which gives salts called hypochlorites), and chloric acid, $HClO_3$ (which gives salts called chlorates). HF gives hydrofluoric acid, HCl gives hydrochloric acid, HBr gives hydrobromic acid, and HI gives hydriodic acid. The equation for the reaction between hydrogen chloride and water is:

$$HCl(g) + H_2O(l) \longrightarrow H_3O^+(aq) + Cl^-(aq)$$
$$\text{(Section 10)} \qquad \text{chloride ion}$$

Hydrogen chloride is therefore very soluble in water. Hydrogen bromide and hydrogen iodide which are more readily ionized are even more soluble. All three produce very strong acids (Section 33). Hydrofluoric acid is comparatively weak; partly because the bonds between hydrogen and fluorine are so strong and partly because of hydrogen bonding (Section 11). The hydrolysis of HF does not therefore take place nearly so readily and at 18 °C a decimolar solution of hydrogen fluoride is only about 10% ionized.

The most important of the halogen hydracids is hydrochloric acid, still sometimes known as 'spirits of salt'. The latter name gives a clue to its usual laboratory preparation – the action between concentrated sulphuric acid and damp salt (Section 75). To prepare the acid, the gas evolved is passed straight into distilled water using an inverted funnel attachment to prevent 'sucking-back' as the gas is so soluble.

ACTION OF THE HALOGENS ON WATER: Fluorine attacks water vigorously producing oxygen gas and a little hydrogen peroxide. Chlorine dissolves to produce 'chlorine-water' which has a pale yellow/green colour. There is some chemical action which is assisted by daylight (photochemical) and chlorine-water does not keep very well. First the chlorine reacts with water to produce hypochlorite ions (reversible reaction):

$$Cl_2 + H_2O \rightleftharpoons 2H^+ + Cl^- + ClO^-$$

This is followed by a slow, irreversible and photochemical decomposition of the hypochlorite ions producing more chloride ions and oxygen gas:

$$2ClO^- \longrightarrow 2Cl^- + O_2(g)$$

Thus the chlorine-water, if kept for a day or two, becomes colourless dilute hydrochloric acid. Chlorine-water or damp chlorine gas bleaches because of the formation of hypochlorite ions (Section 38). Bromine also dissolves in water to give 'bromine-water' and this keeps better because the reaction with water is much slower. Its strength can be maintained by keeping a few cm^3 of liquid bromine in the bottom of the bottle. Its reactions with water are similar to those of chlorine but much slower. Bromine-water contains the ions H^+, Br^-, BrO^- (hypobromite). This dark-brown solution also bleaches because of the presence of hypobromite ions. Iodine dissolves very slightly in water – just enough to colour the solution pale yellow – but no chemical action is evident.

HALOGEN OXIDES: All the elements can form oxides but, as might be expected from their relative positions in the activity series, these are generally unstable. Chlorine, bromine, and iodine each form more than one oxide and, rather strangely, the oxides of bromine have been more difficult to obtain than those of chlorine. The most stable of all the oxides is an oxide of iodine, I_2O_5.

ACTION WITH METALS: These elements attack all metals forming, by direct synthesis, the metallic halides (fluorides, chlorides, bromides, iodides). Chlorine rapidly attacks both gold and platinum. The vigour of the reactions diminishes from fluorine to iodine.

OXIDIZING POWER: Fluorine is the most powerful oxidizing agent and iodine the least. It is because of this that chlorine can, in aqueous solution, displace bromine from a bromide, or iodine from an iodide. Bromine can displace only iodine from an iodide, whilst iodine cannot displace any of the other three halogens:

$$Cl_2(g) + 2Br^-(aq) \longrightarrow 2Cl^-(aq) + Br_2(l)$$
$$Cl_2(g) + 2I^-(aq) \longrightarrow 2Cl^-(aq) + I_2(s)$$

$$Br_2(l) + 2I^-(aq) \longrightarrow 2Br^-(aq) + I_2(s)$$

Methods of obtaining the elements

Fluorine is obtained by the electrolysis of a fluoride melt. Dry hydrogen fluoride, being covalent, does not conduct but when it is mixed with pure potassium fluoride, KF, a conducting addition compound, KF.HF, is formed. The cell, made of copper or steel, is at once attacked by the fluorine but this forms a protective layer on the metal surface. The anode is made of graphite.

Chlorine can be isolated by the oxidation of hydrochloric acid, but most of it is produced as a by-product of the manufacture of sodium hydroxide (Section 38) and the extraction of sodium metal (Section 39).

Bromine comes from the bromides contained in sea water (mainly potassium and magnesium) which are present in small quantities. The bromine is released from the compounds by oxidation with chlorine gas.

Iodine is obtained mainly from sodium iodate, $NaIO_3$, which is found in the Chilean nitrate deposits. The crude nitrate contains about 0.1% of iodine by mass and this is extracted by a rather complicated process.

In elementary work, chlorine and iodine are the more important halogens to study and further information about these is given below.

Chlorine

This gas was discovered in 1774 by the Swedish chemist, Scheele, when he heated spirits of salt (hydrochloric acid) with pyrolusite (a natural form of manganese(IV) oxide, MnO_2). The method became an industrial one but was never very efficient. Manganese(IV) oxide can act as an oxidizing agent and we now know that any oxidizing agent will liberate chlorine from hydrochloric acid – the reaction is, in fact, used as a test for an oxidizing agent. The cheapest oxidizing agent is air and Deacon, in 1868, founded an industrial method which employed a catalyst to enable air to liberate the chlorine from hydrogen chloride. The yield, however, was never more than 10% by mass of the chlorine available. In modern times, by using a rare transition metal as catalyst and providing the right conditions, the Shell Petroleum Co. have revived the method, raising the percentage yield to between 80% and 90%. In the laboratory, the oxidizing agent chosen is now usually potassium manganate(VII) (permanganate) and the preparation is carried out in a fume chamber (Section 75).

If concentrated hydrochloric acid is electrolysed in a voltameter fitted with graphite electrodes, hydrogen forms at the cathode and chlorine at the anode. One would expect the two gases to be evolved in equal volumes ($2HCl \longrightarrow H_2 + Cl_2$) but, because chlorine dissolves in the acid, it is some time before much of this gas is collected. If the acid is thoroughly saturated with chlorine before the electrolysis, the two gases can be seen to be evolved in equal volumes.

Chlorine gas can be converted from molecules to chloride ions either by direct combination with a metal such as sodium or by combining the gas with hydrogen and adding the covalent hydrogen chloride to water. Chloride ions can be converted to chlorine gas (molecular) either by electrolysis of a chloride solution using graphite electrodes or by mixing a chloride (solid) with manganese(IV) oxide and gently warming the mixture with concentrated sulphuric acid. The hydrogen chloride released from the chloride by the acid is oxidized by the manganese(IV) oxide to chlorine gas:

$$MnO_2(s) + 2NaCl(s) + 3H_2SO_4(l) \longrightarrow MnSO_4(aq) + 2NaHSO_4(aq) + 2H_2O(l) + Cl_2(g)$$

This is a general test for the presence of a chloride and the equation can be remembered as the '1,2,3, equation' – 1 mole of the oxide, 2 of a monovalent metal chloride, 3 of the acid (present in excess).

If chlorine is passed over cold, dry calcium hydroxide (slaked-lime) 'bleaching powder' is formed. It consists of a mixture of calcium hypochlorite, $Ca(ClO)_2$, and the basic chloride (Section 37) of calcium. The powder provides a convenient way of storing chlorine in solid form, because any dilute acid (even carbonic) will liberate chlorine from it. It does not keep very well. The importance of bleaching powder has greatly decreased now that liquid chlorine is readily available stored in cylinders made from special steel. A leak from one of these cylinders is extremely dangerous because $1 cm^3$ of the liquid gives about $460 cm^3$ of the gas if it vaporizes at room temperature and pressure.

Although chlorine explodes with hydrogen if the mixed gases are ignited, a burning jet of hydrogen will continue to burn quietly if lowered into a jar of chlorine forming hydrogen chloride. It is essential that all air is swept out of the hydrogen-producing apparatus before the jet is lit, as a mixture of air and hydrogen also explodes.

Industrially, if chlorine is overproduced and there is a greater demand for hydrochloric acid, a conversion from one to the other can be carried out in this way:

$$H_2(g) + Cl_2(g) \longrightarrow 2HCl(g)$$

The explosion of hydrogen with chlorine is most violent when the volumes are equal.

Chlorine reacts slowly with alkanes giving substitution products (Section 53) but vigorously with unsaturated hydrocarbons (Section 54).

Chlorine has many uses. It is a powerful bleaching agent when damp though, for many purposes, a milder agent such as hydrogen peroxide or sulphur dioxide is preferred. As a powerful germicide it is used for sterilizing drinking water. Many organic compounds such as trichloroethylene, benzene hexa-chloride ('gammexane'), polyvinyl chloride (PVC), refrigerants, fire extin-guishers, dyes, and explosives require chlorine in their manufacture.

Iodine

The pure element is a black, shiny solid which, when heated, forms a violet vapour. It both sublimes and dissociates into atoms. In everyday life, iodine is known in solution as 'tincture of iodine'. As iodine is almost insoluble in water, tincture of iodine is made in a special way. Iodine is readily soluble in organic solvents such as carbon disulphide, trichloromethane, and tetra-chloromethane giving a violet-coloured solution in each case (the same colour as iodine vapour). The molecular state of iodine is believed to be I_2 in these solvents. In solvents whose molecules contain oxygen, however, the solution has a dark-brown colour and this is due to a linkage between iodine and the solvent molecules (**solvation**). This occurs with ethanol, ethoxyethane ('ether'), and water (slight solubility). To dissolve iodine without using an organic solvent a concentrated solution of potassium iodide can be used (but this is rather expensive). The molecules of iodine react with iodide ions to give com-plex ions of I_3^-:

$$I_2(s) + I^-(aq) \longrightarrow [I_3]^-(aq)$$

Tincture of iodine is prepared by dissolving iodine (at various concentra-tions) in aqueous potassium iodide with ethanol as well. Although this solu-tion is a good antiseptic it is rather crude in its action, killing useful cells as well as bacteria. Like the other halogens, iodine is very poisonous and the application of too much tincture can produce blood poisoning. The dark-brown solution of iodine used in the laboratory is prepared by dissolving the solid element in aqueous potassium iodide solution.

When added to a starch dispersion, (Section 28), a dilute solution of iodine gives a characteristic blue colour. This is a sensitive test, but both solutions should be very dilute otherwise the colour is so intense as to be almost black. When heated, the colour disappears only to return again when the mixture is cooled. The nature of the blue compound is not fully known.

Oxidizing agents liberate iodine from an acidified solution of potassium iodide. A brown colour immediately appears and the iodine can be confirmed by diluting and testing with starch dispersion. 'Starch-iodide' papers, used for testing oxidizing agents, have been prepared by soaking the paper in a mixture of potassium iodide solution and starch dispersion and then drying it. Any oxidizing agent will liberate iodine which immediately reacts with starch to turn the paper blue.

A little iodine is essential for health in the human system and lack of it can lead to the disease of goitre (enlargement of the thyroid gland).

24   Families of Elements: The Noble Gases

This group can be regarded as Group O, indicating that the atoms of these elements have no valency electrons, or as Group VIII, indicating that the atoms have their full complement of electrons. For many years, these elements were known as the 'rare-gas' group but, as argon constitutes nearly 1% of the air by volume, the name is not very appropriate. Later, they became known as the 'inert-gas' group, but when compounds were formed in the case of two or three of them, this name also was dropped. Nevertheless, the gases *are* far more inert chemically than any other elements. The term 'noble gases', which is now the usual one for them, refers to the fact that they are all monatomic, having atoms with valence shells containing complete octets of electrons (except helium which has only the first shell completed with two electrons). The specific heat ratio for each gas was found to be 1.66 (Section 3). As the elements have such low chemical activity, there is little to learn about them other than their history and uses.

The first noble gas to be isolated was argon. Ramsay remembered that Cavendish had written in 1785 'that if there is any part of the phlogisticated air [nitrogen] of our atmosphere which differs from the rest .... we may safely conclude that it is not more than $\frac{1}{120}$th part of the whole'. Therefore Ramsay, in 1894, set out to discover the identity of this $\frac{1}{120}$th part. In the same year, Rayleigh, a well-known physicist, when measuring very accurately the relative densities of various gases, had discovered that samples of nitrogen taken from the air gave values that were about 0.5% higher than those obtained from pure chemical compounds such as ammonia or ammonium nitrite. This made Ramsay suspect there must be a small amount of a gas in the air that was denser than nitrogen (the relative density of argon is 20 compared with that of nitrogen, 14).

The apparatus Ramsay used to isolate the new gas is still preserved in the Science Museum, London. After removing carbon dioxide, water vapour, and oxygen from the air, he passed the remaining gas over red-hot magnesium to remove the nitrogen in the form of magnesium nitride, Mg_3N_2. Nearly 1% of the air by volume still came through all the reagents without change. Ramsay named the new gas 'argon' derived from the Greek word meaning 'idle' (inactive).

About 1868 it had been observed that the spectrum obtained from the Sun's chromosphere showed the presence of an element (seen in the yellow part of the spectrum near the sodium D-lines) not known on Earth. Later it was discovered that pitchblende (a mineral containing uranium oxide, U_3O_8) and cleveite (a complicated mineral containing uranium and thorium) held a gas (formed by radioactivity) which gave the same lines in the spectrum as the unknown Sun-gas. In 1897, Ramsay and Travers obtained this gas from the atmosphere and adopted a name for it that had been suggested earlier – 'helium' from the Greek 'helios' meaning 'the Sun'. The gas was found to have only twice the density of hydrogen.

Ramsay now had two new gaseous elements that had no place in the Periodic Classification and he suspected that there might be a whole new group to be added to a classification that had been regarded as complete. He therefore set himself to find perhaps three or four missing members of the group. To do this he liquefied large quantities of argon obtained from the air. (Very fortunately, during the 1890s, physicists had invented means for liquefying the more difficult gases.) By fractional distillation of the liquefied argon, Ramsay was able to separate and identify (1898) three more inert gases (Table 9).

Table 9

Gas		Relative density
Neon	(Greek 'new')	10
Krypton	(Greek 'the hidden one')	42
Xenon	(Greek 'the stranger')	65+

All these were shown to be monatomic elements and to be present in the air in minute quantities.

The last member of the group had been known as an 'emanation' given out during the radioactive disintegration of radium. It was recognized by

Table 10

Gas		Atomic number (Z)	Boiling point/°C
Helium	He	2	−269
Neon	Ne	10	−246
Argon	Ar	18	−186
Krypton	Kr	36	−153
Xenon	Xe	54	−108
Radon	Rn	86	−62

Ramsay, early in the twentieth century, as the last and densest of the rare gases. It was first called 'niton' but the name 'radon' was eventually adopted.

All the gases are insoluble in water. The boiling point of each rises successively as the atomic number gets greater (Table 10).

This is to be expected as an element with molecules (atoms) of greater mass will have a lower vapour pressure, at the same temperature, than one whose molecules are less sluggish. Therefore the vapour pressure will equal atmospheric pressure at a higher temperature.

Compounds formed by the noble gases

For more than fifty years following their discovery, chemists believed that these gases were totally inert and were not capable of forming compounds with any other element. It came as a shock to most scientists, therefore, when in 1962 a complex salt of xenon with platinum and fluorine, $Xe^+PtF_6^-$, was isolated and, shortly after this, xenon tetrafluoride, XeF_4, was obtained by direct combination of the elements when passed over nickel (as catalyst) at dull-red heat. The fluoride was obtained as white crystals that were reasonably stable when dry. Hydrolysis, however, produced the explosive oxide, XeO_3. Other compounds soon followed: the relatively simple ones XeF_2, XeF_6, KrF_2, KrF_4, RnF_4, and many more complex salts. As the noble gases have atoms with no valence electrons, such compound formation is not easy to explain. It is noteworthy, however, that compounds have been formed only with very electronegative elements – oxygen and fluorine – and only, so far, by noble gases of high atomic number.

Uses of the noble gases

As these gases are so inert, their important uses depend almost entirely on this property. Argon, being the most plentiful is also the cheapest. The others are much more expensive to produce and only find small-scale uses.

HELIUM :   Airships and balloons. A mixture of hydrogen and helium can be made that is just non-flammable and has almost the lifting power of hydrogen itself.

Divers' breathing-apparatus and submersible work-chambers. A helium/oxygen mixture is much better than air because helium under pressure is not absorbed into the blood stream as much as the nitrogen in air is.

Low-temperature research and also as coolant in nuclear reactors. Helium is mainly obtained from natural gas occurring in some parts of the U.S.A. The gas can contain 0.5–2% by volume of helium. In England, the King's Well at Bath contains 0.15% helium.

NEON :   Advertisement signs. The bright-red glow obtained when an electric discharge is passed through a near-vacuum tube containing a little neon was first noted by Ramsay and his co-workers and from 1910 onwards was exploited for use in advertising. Other colours are obtained by using argon, helium, mercury vapour, etc., sometimes contained in coloured glass tubes. Since 1933, the introduction of fluorescent compounds has greatly increased the range of colours available.

Low-temperature research.

Street lighting, in association with mercury-vapour or sodium-vapour lamps.

(The neon ionizes first and the red glow is noticeable before the lamp comes into full operation.)

Various indicator lamps.

ARGON: For gas-filled electric lamps, usually mixed with nitrogen. The gas allows the tungsten filament to be raised to a high temperature without deterioration.

Argon-arc welding. Provides an inert atmosphere. When metals are welded, formation of any oxide film on the metal surface prevents a good join, particularly in the case of aluminium.

Providing an inert atmosphere when the metal titanium is obtained by reducing titanium(IV) chloride with sodium.

In the manufacture of Geiger counters.

KRYPTON: Miners' cap-lamps. Small projector-lamps. Iodine/quartz lamps for car headlights.

XENON: For highly efficient high-pressure short-arc lamps. Special lamps for lighthouses.

RADON: Radioactive. Has had some use in the treatment of skin cancer. It emits alpha-particles and has a half-life of 3.8 days only.

25 Hydrogen. Hydrides

HYDROGEN

Hydrogen realizes the age-old concept of a fundamental material. Astronomers know that nebulae consist of this element and, out of this must have come atoms of all the other elements. An atom of hydrogen has the simplest structure – one proton with one electron – whereas atoms of all other elements contain two or more protons with neutrons and accompanying electrons.

In the activity series (Section 6) hydrogen comes about half-way down the list of elements. It is not, therefore, strongly electropositive or strongly electronegative. It combines readily only with the elements at the extreme ends of the list and forms more stable compounds with those at the electronegative end. It combines directly with metallic elements down as far as magnesium and these hydrides are ionic – hydrogen being the negative ion. At the other end of the activity series, hydrogen forms, by direct combination, stable covalent compounds with chlorine, oxygen, and fluorine. It forms two oxides – water, which is neutral, and hydrogen peroxide, H_2O_2 (Section 32), which is weakly acidic. When aqueous solutions are electrolyzed, if hydrogen is evolved it always appears, like metals, at the cathode. On the other hand, if fused lithium hydride, Li^+H^-, is electrolysed, hydrogen gas is evolved at the anode. Hydrogen is intimately connected with acids (Section 33).

Considering its fundamental nature, it is appropriate that hydrogen was the first gaseous element to be isolated – by Cavendish in 1766. He obtained it by the action of either dilute vitriolic (sulphuric) acid or spirits of salt (hydrochloric acid) on zinc, iron or tin. Hydrogen is still prepared in the laboratory by Cavendish's method (Section 75). Nitric acid cannot be used because its strong oxidizing action produces water instead of hydrogen, the acid itself being reduced to oxides of nitrogen (Section 64).

For the laboratory preparation of hydrogen, 'granulated' zinc is usually used with the acid. This form of zinc is prepared by pouring molten zinc into cold water. It presents a larger surface area for reaction with the acid. If small quantities of other less electropositive metals (or graphite) are present tiny electrolytic cells are set up and zinc, being the negative pole, goes into solution more quickly. Thus if the reaction between zinc and acid is too slow, the addition of a few drops of copper(II) sulphate solution produces a little copper and the reaction noticeably quickens.

The industrial preparation is now usually from methane or the light naphtha fraction from the distillation of petroleum (Section 53) in a catalytic steam/hydrocarbon reforming process (Section 75). At least two-thirds of the world production of hydrogen goes to the manufacture of ammonia (Section 63). Other important uses are in hydrogenation processes (Section 54) and as a rocket fuel.

Hydrogen is commonly supplied in steel cylinders which contain the gas under high pressure. If these have to be transported, the mass of the gas is negligible compared with that of the cylinders and the cost is high. One way of lessening the cost is to transport liquid ammonia and 'crack' this to obtain hydrogen where and when it is needed (Section 63).

Hydrogen is the lightest gas known and its molecules are diatomic (Sections 3 and 9). At very high temperatures, the molecules split into single atoms becoming 'atomic hydrogen':

$$H_2 \longrightarrow 2H \qquad \Delta H = 431\,kJ$$

Because this reaction is so strongly endothermic (Section 46), very large quantities of heat are evolved when the molecules reform and this fact is put to good use in the 'atomic hydrogen torch'. Aluminium oxidizes on the surface very readily, but if the welding of this metal is carried out with such a torch, molecules of hydrogen form on the surface producing a very high temperature whilst, at the same time, shielding the metal from oxidation. Atomic hydrogen is a much more powerful reducing agent (Section 31) than is molecular hydrogen.

Two other isotopes (Section 2) of hydrogen are known: ^2_1H, Deuterium (symbol D) and ^3_1H, Tritium (symbol Tr). Tritium is radioactive.

Molecular hydrogen will reduce a few metal oxides when the oxide is heated in a stream of the gas:

Copper(II) oxide: $CuO(s) + H_2(g) \longrightarrow Cu(s) + H_2O(g)$
Lead(IV) oxide: $PbO_2(s) + H_2(g) \longrightarrow Pb(s) + H_2O(g)$

On an industrial scale metallic tungsten, for the manufacture of electric-lamp filaments, is obtained from tungsten(VI) oxide (trioxide) by heating it in hydrogen at about 800 °C. Tungsten has the symbol W derived from the name of its ore, wolframite:

$$WO_3(s) + 3H_2(g) \longrightarrow W(s) + 3H_2O(g)$$

Hydrogen is a constituent of more compounds than is any other element; principally because of the vast number of organic compounds that contain it.

HYDRIDES

All non-metals (with the exception of the noble gases) and the most electro-positive of the metals can form, under the right conditions, compounds with hydrogen. Table 11 shows some examples listed according to the Periodic Classification:

Table 11

Group	I	II	III	IV	V	VI	VII	O
	LiH			CH_4	NH_3	H_2O	HF	—
	NaH		$(AlH_3)_n$	SiH_4	PH_3	H_2S	HCl	—
	KH	CaH_2					HBr	—
							HI	—

Hydrides of the first two groups are ionic. The others are covalent. Aluminium hydride exists only as a polymer (Section 57). In the case of carbon (Group IV) there are limitless numbers of other hydrides (hydrocarbons) because of the power of **catenation** of the carbon atoms (Section 52). As far as Group IV, the valence of each element has the same figure as the group it is in, but after that it descends from IV to O. When the element combining with hydrogen is much more electronegative (as in Groups VI and VII) the hydrogen symbol is placed before the other. The hydrides of Group IV are gases that are insoluble in water. The hydrides of Group V are gases, basic in nature, but only ammonia dissolves to form an alkali. In Group VI, water is a neutral liquid but hydrogen sulphide dissolves in water to give a weakly acid solution. All the hydrides of Group VII are very soluble in water giving acids.

The hydrides of the elements of Groups I and II are white solids that are easily hydrolysed by water (Section 20). Calcium hydride has been used for the preparation of hydrogen for the filling of small balloons, thus avoiding the transport of heavy cylinders of gas. Hydrogen is evolved from the hydride as soon as water is added:

$$CaH_2(s) + 2H_2O(l) \longrightarrow Ca(OH)_2(s) + 2H_2(g)$$

The solid product is calcium hydroxide (slaked-lime) which is not very soluble in water.

26 Water

At the end of the eighteenth century, the gases then known were all called 'airs': vital air (oxygen), fixed air (carbon dioxide), nitrous air (nitrogen monoxide). Cavendish found that his 'inflammable air' (hydrogen) would burn quietly if fairly pure but if mixed with ordinary air or oxygen (when discovered later) a dangerous explosion could result when the mixture was ignited. Cavendish believed that when the explosion took place a little water was formed (though to make 1 cm^3 of water we now know that more than 2 litres of hydrogen and oxygen in the right proportions are needed). Cavendish found, by exploding fairly large quantities of the mixed gases and identifying the measured volume of gas left over, that the correct proportion for complete

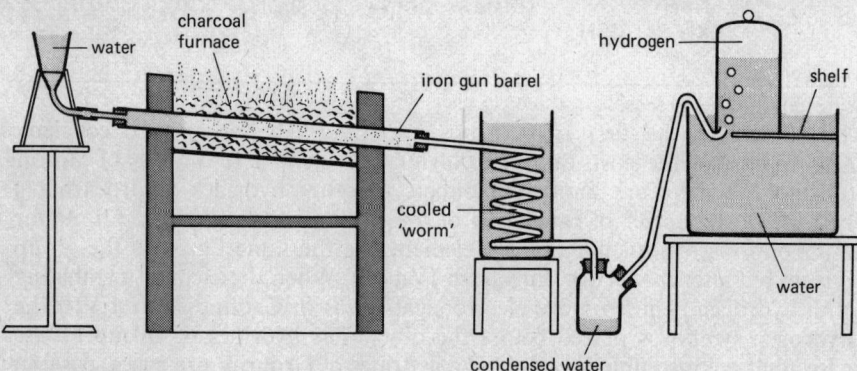

Figure 4 Lavoisier's 'gun-barrel' apparatus for obtaining hydrogen from water (total length about 3 m)

reaction was $2\frac{1}{2}$ air/1 hydrogen or 1 oxygen/2 hydrogen by volume. Unfortunately, the water he obtained by exploding hydrogen with oxygen was found to contain an acid – identified by Cavendish as nitric acid (the oxygen he used could not have been free from nitrogen). Later, Lavoisier repeated the experiments, and using the proportions by volume of 2 hydrogen: 1 oxygen, did obtain fairly pure water.

Lavoisier further argued that if it were possible to obtain water from hydrogen, then it should be possible to obtain hydrogen from water. This led him to prepare the famous 'gun-barrel' experiment in 1784.

An iron gun barrel was made red-hot in a charcoal furnace and water was run from a funnel through the tube (Figure 4). The water first turned into steam and this reacted with the iron. Any unchanged steam was condensed to water in a cooled glass spiral fixed at the end of the barrel and the hydrogen was collected over water. The inside of the barrel was later shown to be coated with what appeared to be shiny black 'smithy-scale' (an iron oxide), but this Lavoisier could not finally prove. Lavoisier also tried a copper tube in place of the iron one and found that no decomposition of the water took place, only distillation.

To prove that water *is* formed when hydrogen burns in air, it is only neces-

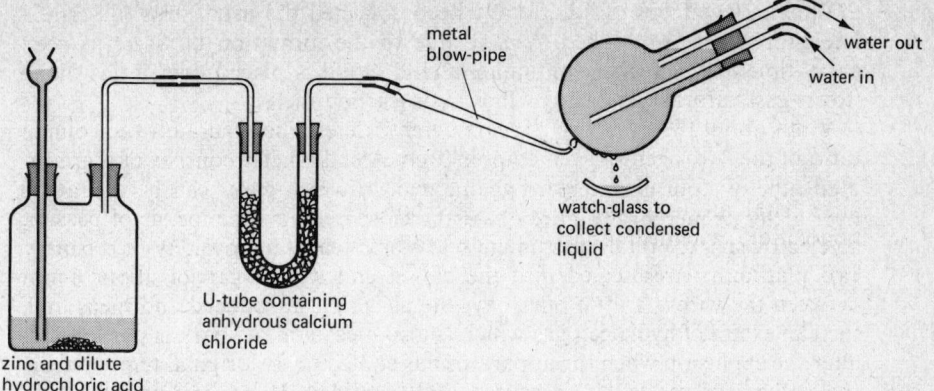

metal
blow-pipe

water out

water in

watch-glass to
collect condensed
liquid

U-tube containing
anhydrous calcium
chloride

zinc and dilute
hydrochloric acid

Figure 5

sary to light a jet of pure hydrogen (dried by passing it through anhydrous calcium chloride) and let it impinge on the surface of a glass flask kept cool by circulating water (Figure 5). (It is very important to make sure that all air has been swept from the apparatus before the jet is lit.)

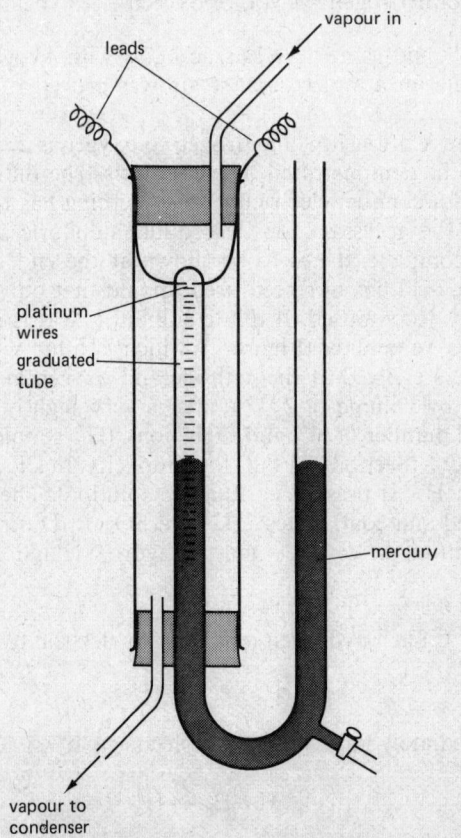

vapour in

leads

platinum
wires

graduated
tube

mercury

vapour to
condenser

Figure 6   A eudiometer

When a few drops of liquid have been collected the usual tests (described later) for water are applied. A mist, due to the formation of water, is seen every time that a beaker containing a cold liquid is placed over a gas flame (town gas, natural gas or any other hydrocarbon fuel).

Volta, about 1790, first used a **eudiometer** to determine accurately the volume ratio of the hydrogen/oxygen combination. A eudiometer consists of a graduated tube of stout glass, closed at one end, in which gases can be contained over water or mercury (Figure 6). At the closed end is ome means of passing an electric spark. Volta's instrument had a brass cap, but nowadays it is simply two platinum wires sealed into the closed end with a gap of about 1mm between the wires. A little pure oxygen gas is first introduced and measured, then an excess of hydrogen gas which is also measured. A spark is passed and, after the explosion when the apparatus has cooled to the original temperature, the volume of gas left is measured and identified. Hence the volume ratio can be found. If the whole tube is surrounded by the vapour of some liquid that has a boiling point well above 100 °C (e.g. amyl alcohol, b.p. 137 °C) the water formed remains, not as a negligible drop but as a measurable volume of steam.

With such an apparatus the volume ratio was found to be (all measurements being made at the same temperature and pressure):

$$\text{2 vols of hydrogen} + \text{1 vol. of oxygen} \longrightarrow \text{2 vols of steam}$$

From this result, and using Gay-Lussac's Law with Avogadro's hypothesis, the correct formula for a molecule of steam was proved to be H_2O (Section 3).

That the volume relationship of hydrogen to oxygen is $2:1$ in the formation of water can also be demonstrated by electrolysis. The difficulty here is that pure water does not conduct electricity; so something has to be added to the water to provide the necessary ions, e.g. dilute sulphuric acid. Therefore, to make the proof complete, it has to be shown at the end of the experiment that the sulphuric acid has not been used up and that only a small quantity of water has been decomposed. If dilute sulphuric acid is electrolysed in an apparatus called a **voltameter** (Figure 7), which is fitted with platinum electrodes, hydrogen is evolved at the cathode and oxygen at the anode in the exact proportion by volume of $2:1$. Water is very slightly ionized and contains a very small number of H^+ and OH^- ions. (H^+ should more accurately be written as H_3O^+ (Section 10) but, for simplicity, this fact is always taken for granted when H^+ is present in aqueous solution). The sulphuric acid is completely ionized and contributes $2H^+$ and SO_4^{2-}. During electrolysis, the hydrogen ions drift to the cathode and the hydroxide and sulphate ions drift to the anode.

At the cathode $(-)$: Each hydrogen ion gains an electron to become an atom.

$$H^+ + e^- \longrightarrow H$$

Two atoms immediately pair to give a molecule:

$$H + H \longrightarrow H_2$$

Hydrogen molecules are discharged as hydrogen gas.

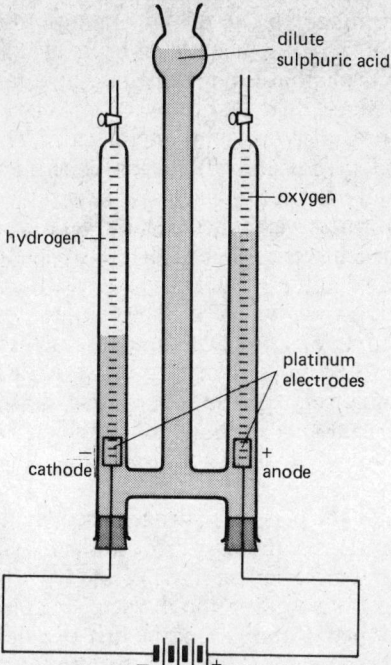

Figure 7 A voltameter

At the anode $(+)$: The hydroxide ions, although in much smaller concentration, are discharged preferentially losing their electrons.

$$OH^- - e^- \longrightarrow OH$$

A hydroxyl group does not exist in a free state and two hydroxyl groups immediately react to form water and an atom of oxygen.

$$OH + OH \longrightarrow H_2O + O$$

Oxygen atoms then pair to form molecules and the gas is discharged. As hydrogen and oxygen gases are discharged at the electrodes, more water molecules dissociate to maintain the original equilibrium.

To sum up:

$$2H^+ 2OH^- + (2H^+ + SO_4^{2-}) \longrightarrow 2H_2 + O_2(2H^+ + SO_4^{2-})$$

The composition of water by mass set a rather more difficult problem. In 1842 Dumas made use of the reducing action of hydrogen on heated copper(II) oxide:

$$CuO + H_2 \longrightarrow Cu + H_2O$$

In order to minimize error he carried out his experiments (nineteen in all) on a very large scale. He purified the hydrogen by passing it through seven

U-tubes containing agents to absorb the known impurities and each was over a metre high! An error of 0.01 g in 1 g is an error of 1% whereas an error of 0.01 g in a kilogram is negligible. (So greatly have apparatus and experimental techniques improved since then that, nowadays, very accurate quantitative work is performed on a micro or semi-micro scale.) The copper oxide tube was carefully weighed before and after each experiment, the loss of mass representing the mass of oxygen that had combined with the hydrogen passing through. The water formed was collected in a vessel followed by four large drying-tubes, all of which were weighed before and after the experiment in order to find the mass of water formed. The mass of the hydrogen was finally obtained by subtracting the mass of the oxygen lost by the oxide from the mass of water collected. In any one experiment, something like 60 g of oxygen was found to form rather more than 67 g of water. Taking the average of the results from his nineteen experiments, Dumas decided that the ratio by mass was 7.98 oxygen: 1.00 hydrogen. Knowing that two atoms of hydrogen combine with one atom of hydrogen to form water, this would have given to oxygen a relative atomic mass of 15.96.

The method of finding the mass of hydrogen used 'by difference' was not for long accepted by chemists, because any error in the mass of the oxygen would lead to a corresponding error in the mass of the hydrogen. Later, therefore, methods were devised for weighing the oxygen, weighing the hydrogen, *and* weighing the water formed. If the sum of the first two did not equal the third, there was obviously some error. Dumas' ratio was found to be a little too high and the new accepted value became 7.94 : 1.00, giving a relative atomic mass for oxygen of 15.88 (on the old scale of $H = 1.00$).

Although water is such a commonplace, everyday compound, to the chemist its properties are of unusual and special interest.

(1) *Although mainly a covalent compound, it has a slight ionization into H^+ and OH^-. Pure water is completely neutral. Its pH is 7* (Section 36) but its slight ionization enables it to act, under particular circumstances, either as a very weak acid or as a very weak alkali (Section 33 and 35). In other words, it is capable of acting as a proton donor or as a proton acceptor and is said to be **amphoteric.**

(2) *The shape of its molecule is angular and the oxygen atom has two 'lone pairs' of electrons.* These enable molecules of water to attach themselves to ions (particularly metal ions) to form hydrates (Section 29).

(3) *Because of hydrogen bonding (Section 11) water has an unusually high boiling point.* The hydride of sulphur (next below oxygen in Group VI) is a gas, H_2S, and its boiling point is $-60\,°C$ (at atmospheric pressure).

(4) *Because of its unusually high dielectric constant, water is an excellent ionizing solvent* (Section 8).

(5) *Chemically it is a fairly reactive compound,* both with elements and with other compounds.
With elements: It reacts vigorously with very electropositive ones and the readiness to react falls off as the element is lower in the activity series (Section 6). The metal replaces the hydrogen to give a base (Section 35):

e.g. $$2Na + 2(H.OH) \longrightarrow 2(Na^+OH^-) + H_2$$

It also reacts with the most electronegative elements such as fluorine, chlorine, and bromine (with diminishing vigour) but here acids are produced. For example, with chlorine and cold water:

$$Cl_2 + 2(H.OH) \longrightarrow \underset{\substack{\text{hypochlorous} \\ \text{acid}}}{H^+ClO^-} + H^+Cl^-$$

With compounds: Many are hydrolysed (Section 20).
(6)   Pure water has the following special physical properties:
    (a)   Its maximum density is $1\,g\,cm^{-3}$ at $4\,°C$.
    (b)   It boils at $100\,°C$ at $760\,mm$ pressure.
    (c)   It freezes at $0\,°C$.
Two at least of these are necessary to identify a liquid as water.

27   Natural Waters. Hardness

NATURAL WATERS

The principal types of water that occur in nature are rain, spring, river, and sea. Rain water falls through the air, percolates through the ground, and some issues as spring water. Springs merge to form rivers and rivers find their way to the sea. None of these waters is pure and the elements and compounds they contain may vary widely.

Rain water

As rain falls through the air, some of the soluble gases in the atmosphere will dissolve in it. Thus rain water may gain oxygen and nitrogen in very small quantities, carbon dioxide, and, in the neighbourhood of industrial towns, sulphur dioxide, etc. Rain will also carry down insoluble matter in suspension, such as dust and soot. Rain water that has been filtered through a fine filter is not very different from distilled water obtained in the laboratory.

Spring water

Rain water in its passage through the soil may take up more gases, lose the solid matter in suspension but gain solid matter in solution including the following ions: Ca^{2+}, Mg^{2+}, SO_4^{2-}, Cl^-, HCO_3^-. Spring water is nearly always clear and sparkling. In some districts it may be unusually rich in less common substances such as iron(II) in the chalybeate waters at Tunbridge Wells and hydrogen sulphide at Harrogate. Sometimes it may emerge at a fairly high temperature as at Bath or Buxton.

River water

As rivers are fed by springs, river water will contain all that spring water contains plus solid matter in suspension stirred up and carried along by its movement. In towns it can be badly polluted with industrial waste and even in the country it can be polluted by excessive use of herbicides, artificial fertilizers, etc. drained from the fields.

Sea water

As seas receive water from rivers, one finds in sea water all the gases and ions that river water contains. In addition to these, there is a remarkable amount, normally about 3% by mass, of sodium chloride. The origin of this salt has never been fully explained. In inland seas which are enclosed and have rapid evaporation, such as the Dead Sea, the proportion can be over 20% by mass.

Tap water

The supply of water to large towns often consists of river water that has been collected in reservoirs, filtered, and treated with a little chlorine to kill harmful organisms. Exposure to the air for some days in the reservoir kills, by oxidation, many harmful bacteria and the final treatment with chlorine makes quite sure. That tap water contains dissolved solid matter can easily be shown by carefully evaporating a small amount to dryness on a watch glass. There is an obvious residue. Distilled water, treated in the same way, leaves no residue. A beaker of tap water will usually give a cloudiness when tested for sulphate or chloride ions (Section 77). The presence of dissolved gases may be illustrated by boiling the water and collecting the expelled gases. Because oxygen is slightly more soluble than nitrogen, air boiled from water contains about one third oxygen by volume.

HARDNESS

Water that is 'hard' does not lather easily with ordinary soap. Soap is a detergent but its composition is quite unlike that of modern synthetic detergents which lather quite well even with hard water. Soap is made by hydrolysing fats by boiling them with alkali (**saponification**) and the product usually consists of the sodium salts of organic acids (Section 56). Unfortunately the magnesium and calcium salts of these acids are insoluble in water, so that if there are calcium and/or magnesium ions in the water the soap is precipitated as a 'scum' consisting of the calcium and/or magnesium salts of the acids. Not until all the Ca^{2+} and Mg^{2+} ions have been removed will the water lather.

The sulphates of calcium and magnesium are fairly common in small quantities in ordinary soil and in some places occur in large quantities. Calcium sulphate as 'gypsum' ($CaSO_4,2H_2O$) or 'anhydrite' (anhydrous $CaSO_4$) is slightly soluble in water, whereas magnesium sulphate, as 'epsomite' ('Epsom salt', $MgSO_4,7H_2O$), which is frequently found associated with gypsum, is freely soluble in water. These substances, therefore, contribute the ions Ca^{2+}, Mg^{2+}, and SO_4^{2-}.

Magnesium carbonate, $MgCO_3$, is often found with calcium carbonate,

$CaCO_3$, but these two compounds are not soluble in water. However, water that contains carbon dioxide has a slow solvent action converting carbonate into hydrogencarbonate (bicarbonate) ions thus:

$$CO_3{}^{2-}(s) + H_2O(l) + CO_2(g) \longrightarrow 2HCO_3{}^-(aq)$$

The hydrogencarbonates of calcium, magnesium, and iron(II) can exist in solution, but they are not known in solid form because, as soon as a solution containing hydrogencarbonate ions is boiled, the above reaction is completely reversed (Section 48):

$$2HCO_3{}^-(aq) \longrightarrow CO_3{}^{2-}(s) + H_2O(l) + CO_2(g)$$

When steam is raised in boilers, the deposit of calcium and magnesium carbonates (often known as 'scale' or 'fur') leads to waste of fuel and shortens the life of the boilers. Engineers find that, if water is very hard, it is better to soften the water before it is boiled. This can be done, reasonably cheaply, in a variety of ways:

Temporary hardness

If the water contains mainly hydrogencarbonate ions, boiling the water will soften it in the way described above – but fuel is costly and the deposited scale is a nuisance. Hardness that can be removed by boiling is referred to as **temporary hardness.**

'Lime-soda' treatment

This is the addition of the calculated quantity of a mixture of calcium and sodium hydroxides. The process is cheap and will remove both calcium and magnesium ions. The former are removed as carbonate from hydrogen-carbonate ions:

$$Ca(HCO_3)_2(aq) + Ca(OH)_2(s) \longrightarrow 2CaCO_3(s) + 2H_2O$$

If too much slaked-lime is added it will of course leave extra calcium ions in solution, thus adding to the hardness. For this reason the amounts added have to be carefully calculated. The reaction above is not effective for magnesium ion removal because in this case the carbonate is too soluble. As the hydroxide of magnesium is less soluble than that of calcium, the magnesium ions are precipitated by the sodium hydroxide:

$$Mg^{2+}(aq) + 2OH^-(aq) \longrightarrow Mg(OH)_2(s)$$

Permanent hardness

This is usually caused by the sulphates (see above) and to some extent the chlorides of calcium and magnesium. Boiling has no effect on these, so they have to be *removed by ion aggregation* (Section 18). If a cheap soluble metal carbonate (usually 'washing soda', $Na_2CO_3, 10H_2O$) is added, calcium and magnesium ions are precipitated as insoluble carbonates:

$$Ca^{2+}(aq) + CO_3^{2-}(aq) \longrightarrow CaCO_3(s)$$
$$Mg^{2+}(aq) + CO_3^{2-}(aq) \longrightarrow MgCO_3(s)$$

Sequestering agent

Calcium and magnesium ions can be immediately removed by a **sequestering agent**. Such an agent virtually locks up these ions in a complex aggregate of particles and prevents them from reacting with soap. The word 'sequester' means to isolate or set apart and a commonly used sequestering agent is sodium metaphosphate polymer, $(NaPO_3)_n$ (Section 66). The material looks like soapflakes and is soluble in water.

Ion-exchange material

By filtering the hard water through a suitable **ion-exchange** material the calcium and magnesium ions can be exchanged for sodium ions that do not cause hardness in water. A common material is 'Permutit(e)' which is a complicated hydrated silicate of aluminium and another metal. Sodium permutite (which may be represented simply as NaP) is a hydrated form of sodium aluminium silicate. It reacts with calcium or magnesium ions thus:

$$2NaP + Ca^{2+} \longrightarrow CaP_2 + 2Na^+$$
$$\text{(or } Mg^{2+}) \quad (MgP_2)$$

The great advantage of this process is that it can be reversed and the permutite material restored. Permutite is not soluble and the water trickles slowly over it in a vertical container and emerges, softened, at the bottom. Eventually the sodium permutite is all converted to the Group II metal permutite and is no longer effective. At this stage, if the material is allowed to stand with strong brine (sodium chloride solution) the above reaction is reversed and sodium permutite reformed. The charge is then flushed with water; thus being made good for a few more weeks use, the time depending on the quantity of water that is put through the regenerated permutite.

De-ionized water

Because of its very slight ionization, the conductivity of pure water is very low: 0.04 siemens at 18 °C. The various other ions in tap water result in its conductivity being much higher than this figure. The greater conductivity comes mainly from dissolved gases and solid material. Even distilled water has an appreciable conductivity.

Nowadays, very pure water may be obtained by making use of prepared ion-exchange resins and these produce 'de-ionized' water. The resin polymers (Section 57) contain particular groups of atoms introduced into complicated organic molecules. Water passed through a cation-exchange resin has all positive metal ions exchanged for hydrogen ions; water passed through an anion-exchange resin has all other negative ions exchanged for hydroxide

ions. The cation-exchange resin has a strong acid cation and the anion-exchange resin has a strong base anion. For example:
Removal of calcium ion:

$$Ca^{2+}SO_4{}^{2-} + 2H^+ \text{ resin} \longrightarrow H_2SO_4 + Ca^{2+} \text{ resin}$$

(This resin can be regenerated eventually by standing with dilute hydrochloric or sulphuric acid.)

Removal of chloride ion:

$$H^+Cl^- + OH^- \text{ resin} \longrightarrow H_2O + Cl^- \text{ resin}$$

(This resin would be regenerated with sodium hydroxide solution.)
By using a 'mixed bed' of both types of resin, water can readily be freed from all foreign ions.

Stalactites and Stalagmites

The deposition of calcium and magnesium carbonates not only causes the 'fur' in kettles and the 'scale' in boilers, but also accounts for the formation of stalactites (growing downward) and stalagmites (growing upward) in caves. If rain water trickles slowly through chalk or limestone in large quantity, it eventually becomes saturated with hydrogencarbonate ions and if this saturated solution drips slowly from the roof of a cave, evaporation causes the building up of deposits. It is a very slow process and a stalagmite may grow only 30 cm high in the course of a thousand years!

28   Solubility. Colloidal State

SOLUBILITY

A **solution** *is made when one substance, the* **solute,** *is dispersed in ionic, molecular (or near-molecular) form throughout another substance, the* **solvent,** *without chemical reaction taking place.*

All gases are completely miscible, but it is possible to have solutions of gas/liquid; liquid/liquid; liquid/solid; solid/liquid; solid/solid; the solute being put first in each pair. Some of these systems are uncommon, but solid/solid ('solid solution') is quite common in various metal alloys (Section 44).

Water is the commonest solvent and it will dissolve a large number of inorganic compounds. In organic chemistry (Sections 52–60), water is of less importance as a solvent and solvents such as ethanol and ethoxyethane (ether) are more commonly used. Metals are not soluble in water unless they react with it and the solid non-metallic elements phosphorus, sulphur, and iodine are almost insoluble in water, though each dissolves readily in carbon disulphide, CS_2.

Gases

Normally, gases are almost insoluble in water. Where there is moderate or great solubility, it is a sign that there has been some chemical reaction with the water; so that to say that the gas has 'dissolved' in the water is not strictly correct. The gas has reacted with it.

The solubility of a gas is expressed as the number of volumes that will 'dissolve' in one volume of the solvent at stated temperature and pressure.

The three most soluble of the common gases in water (at s.t.p.) are ammonia (1300), hydrogen chloride (500), and sulphur dioxide (80).

Ammonia: This covalent gas reacts with covalent water to give ammonium and hydroxide ions. Undissociated NH_4OH probably does not exist:

$$NH_3 + H_2O \longrightarrow NH_4^+ + OH^-$$

Ammonia in solution acts only as a weak base and does not appear to be highly dissociated. The great solubility is therefore explained by the formation of hydrates of ammonia by hydrogen bonding (Section 11).

Hydrogen chloride: This covalent gas reacts with water to give oxonium and chloride ions, resulting in the formation of a strong acid (Section 33):

$$HCl + H_2O \longrightarrow H_3O^+ + Cl^-$$

Sulphur dioxide: This, another covalent gas, reacts with water to give sulphurous acid, H_2SO_3, which cannot be isolated. Sulphurous acid is dibasic (Section 33) and can give hydrogensulphite and sulphite ions:

$$SO_2 + H_2O \rightleftharpoons H_2SO_3 \rightleftharpoons H^+ + HSO_3^- \rightleftharpoons 2H^+ + SO_3^{2-}$$

Carbon dioxide: Is not so soluble in water (1.7 vol. to 1 at s.t.p.). It forms carbonic acid which, again, cannot be isolated. It is dibasic giving hydrogencarbonate and carbonate ions (compare SO_2 above):

$$CO_2 + H_2O \rightleftharpoons H_2CO_3 \rightleftharpoons H^+ + HCO_3^- \rightleftharpoons 2H^+ + CO_3^{2-}$$

Gases that are not very soluble act according to *Henry's Law* which states that

The mass of a gas dissolved by a given volume of liquid at constant tempera-ture is proportional to the pressure of the gas, provided that no chemical action occurs.

The solubility of any gas in water decreases rapidly with rise in temperature. If there is no chemical action, the molecules of gas absorbed are jostled out by the increased molecular movement. Where there has been chemical union, the reaction is reversed and the gas expelled. For example, an ammonia solution gives back ammonia gas when the solution is heated; concentrated hydrochloric acid evolves hydrogen chloride gas when heated; sulphurous and carbonic acids also release their gases on heating.

Liquids

Scientists do not talk of liquids being soluble in one another but of their being **miscible, immiscible** or **partially miscible.** If they are immiscible they can easily be separated by a tap funnel (Section 14). Examples of each:
Miscible: Water/ethanol; water/ethanoic (acetic) acid.
Immiscible: Water/carbon disulphide; water/tetrachloromethane.
Partially miscible: Water/ethoxyethane (ether); water/nicotine.

Solids

Electrovalent compounds have their ions dispersed by water lessening the attractive forces between the ions and then distributing the ions by its own molecular movement. Non-ionic compounds which are soluble in water, e.g. cane-sugar (sucrose), are reduced to molecular or near-molecular condition by bombardment of the water molecules. To hasten solution two simple things can be done. First, one can break up the lump of sugar with a rod, thus increasing the surface area exposed to the bombardment of the water molecules; second, one can raise the temperature of the water. This increases the velocity and hence the momentum of the molecules so that they can disintegrate the sugar particles more effectively. In the process of solution, the molecules of solvent lose energy and the heat of solution of any substance has a positive value unless there is chemical action between substance and solvent (Section 46).

In the case of a solid dissolving in water,

> *the solubility of the solid is defined as the maximum mass of solute that will dissolve (excess of solute being present) in 100 g water at a stated temperature.*

The words 'excess of solute being present' have to be inserted because a few solutions are capable of holding, temporarily, more than their saturation figure and so become **'supersaturated'.** A supersaturated solution is very unstable and if just one or two crystals of the solute are added, the excess is immediately crystallized. Example: hydrated sodium thiosulphate, $Na_2S_2O_3,5H_2O$, ('hypo').

Nearly all solids become more soluble in water as the temperature is increased because, as Le Chatelier's principle (Section 48) predicts, if the temperature rises the system will move in the direction that absorbs heat. However, the variation among compounds of the effect of rising temperature is very great. A given mass of water will dissolve very little more sodium chloride at $100\,°C$ than it will at $0\,°C$. With potassium nitrate, on the other hand, there is an enormous difference. 100 g water at $20\,°C$ will dissolve about 30 g of this nitrate. By $70\,°C$, it will dissolve about 150 g! By determining the solubility of a pure compound at a number of temperatures, it is possible to draw a 'solubility curve', from which the solubility can be read at a glance at intermediate temperatures (see Figure 8).

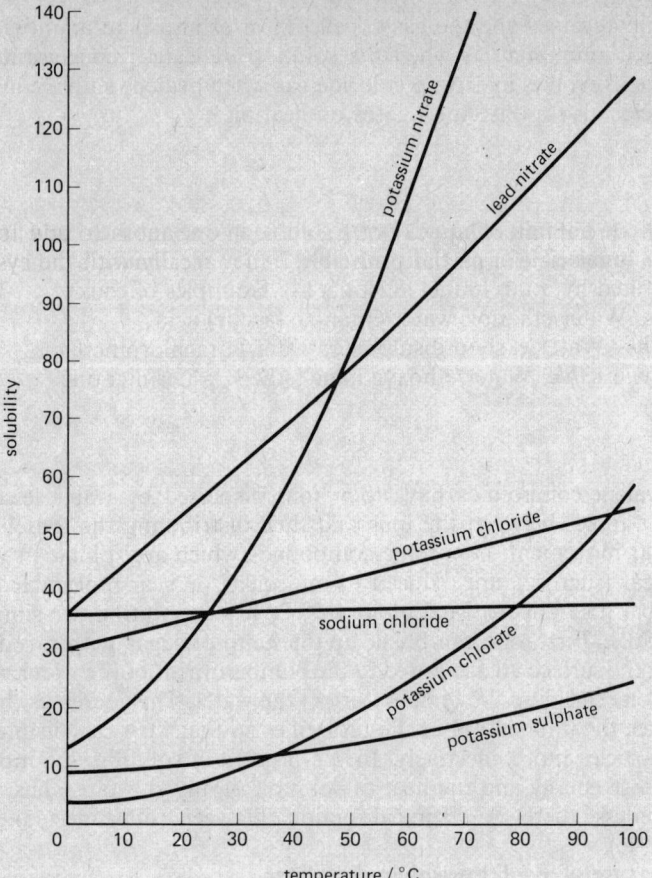

Figure 8   Solubility curves of various pure compounds

COLLOIDAL STATE

If insoluble solid particles, large enough to see, are dispersed throughout a liquid, the system is known as a **suspension.** The particles can easily be separated from the liquid by filtration or by the use of a centrifuge and, if the suspension is allowed to stand, the particles soon settle, unless they are very fine, to form a precipitate.

If the particles dispersed are much too small to see, possibly ions, molecules or small groups of molecules, they cannot be separated from the liquid by any means of filtration and can only be recovered by evaporation of the solvent. Such a system is, of course, a **solution.**

In the colloidal state, which lies between these two extremes, there can be *particles of such a size that they are not large enough to settle under gravity, can only be seen with a specially arranged high-powered microscope, and can readily pass through any ordinary filter paper.* Sometimes the system looks quite clear to the naked eye, but sometimes it has a cloudy appearance (e.g. colloidal sulphur).

Colloids have properties that are quite unlike those of solutions or suspensions. *Colloidal particles are in constant random motion* caused by collisions with molecules of the medium. This random movement was first observed by the botanist Robert Brown, in 1827, and is referred to as 'Brownian movement'.

Colloidal particles carry electric charges which, for a particular colloid, are either positive or negative. The electric charges cause the particles to repel one another and stop them from coagulating into larger units. If the charges can be neutralized the particles soon grow larger, form an ordinary suspension and eventually precipitate. One way of bringing this about is to add ions which carry the opposite charge but do not react chemically with the system. The higher the value of the valence of the ion, the more effective it will be. For example, colloidal iron(III) hydroxide is a positive colloid. It can be converted to an ordinary suspension by adding chloride ions $(-)$, sulphate ions $(2-)$ or orthophosphate ions $(3-)$ (Section 66), and the coagulation is most rapid when the last is added.

Colloids are *not* a special kind of matter and it is now known that any element or compound can be brought into the colloidal state if enough trouble is taken – even sodium chloride. Colloids are very common in nature, e.g. smoke, fog, milk, etc. Gases, liquids, and solids may be involved. Smoke has fine carbon particles dispersed over air; fog has fine water droplets dispersed over air; milk has fats, etc., dispersed over water; good ruby glass has colloidal gold dispersed throughout the glass. In these examples, the carbon, water, fat, and gold are referred to as the **disperse phase** and the air, water or glass are referred to as the **disperse medium.** Further examples will be encountered in other parts of this book. Colloidal graphite, for example, (Section 49) is a very important lubricant. Colloids can be a nuisance to the chemist but quite often the colloidal state can be put to good practical use.

---

29 Hydrates

The shape of the molecule of water is angular and the oxygen atom has two 'lone pairs' of electrons (Section 10). Either of these pairs will allow a water molecule to attach itself by co-ordinate linkage to an ion. Electrons being negative, it is natural that water molecules should attach themselves to a positive (metal) ion, but there are two common cases known where attachment is to a negative ion: one molecule of water to a sulphate ion, SO_4^{2-}, and two molecules of water to an oxalate ion, $(COO)_2^{2-}$. The co-ordinate linkage produces a **hydrate** and the bond is often indicated by an arrow, $\longrightarrow$, indicating the transfer of an electron pair to the recipient ion. As water molecules are neutral, the addition or subtraction of water of hydration in no way affects the charge of the ion itself. Some common divalent metal ions such as Mg^{2+}, Zn^{2+}, Fe^{2+} can accept six water molecules and six is known as their **co-ordination number.** The copper ion, Cu^{2+}, has, rather strangely, a co-ordination number of 4. The hydrated copper(II) ion is shown as Cu surrounded by four water molecules forming a **complex ion,** which is indicated by square brackets.

Hydrated copper(II) sulphate, $CuSO_4, 5H_2O$, is a blue crystalline compound which, when heated gently, loses its colour to form white anhydrous copper(II) sulphate. When cool, if water is added the colour returns and heat is evolved, indicating chemical linkage. It has been observed that, on heating the crystals, not all the water comes off at the same temperature. At about $100\,°C$ the water molecules associated with the copper ions are released and at about $250\,°C$ the other molecule (which may act as a bridge between the copper(II) and sulphate ions) is released. At $750\,°C$ the sulphate decomposes into copper(II) oxide and sulphur trioxide. A detailed way of expressing the composition of hydrated copper(II) sulphate is:

$$\begin{bmatrix} H_2O & & OH_2 \\ & Cu & \\ H_2O & & OH_2 \end{bmatrix}^{2+} \qquad [H_2O, SO_4]^{2-}$$

Molecules of ammonia each have a lone pair of electrons (Section 10) and under certain conditions molecules of ammonia can replace the water molecules forming a copper(II) tetrammine complex (Section 72):

$$\begin{bmatrix} H_3N & & NH_3 \\ & Cu & \\ H_3N & & NH_3 \end{bmatrix}^{2+} \qquad [H_2O, SO_4]^{2-}$$

The water contained in hydrates is often referred to as **water of crystallization.** If a hydrated compound is heated, the crystalline shape is lost when the water is removed.
Examples of well-known hydrated compounds:

Barium chloride	$BaCl_2, 2H_2O$
Calcium sulphate (gypsum)	$CaSO_4, 2H_2O$
Magnesium chloride	$MgCl_2, 6H_2O$
Zinc sulphate	$ZnSO_4, 7H_2O$
Iron(II) sulphate	$FeSO_4, 7H_2O$
Sodium carbonate	$Na_2CO_3, 10H_2O$
Sodium sulphate	$Na_2SO_4, 10H_2O$
Potash alum	$K_2SO_4, Al_2(SO_4)_3, 24H_2O$

Where the number of co-ordinated water molecules is higher than seven it is likely that water molecules play an integral part in the crystal structure.

The colour associated with some hydrated ions is lost or changed when the water is removed, e.g. green iron(II) sulphate, blue copper(II) sulphate losing their colour. Cobalt(II) chloride, $CoCl_2, 6H_2O$, has a pink colour in dilute solution. When heated it becomes anhydrous and the colour changes to blue. This has been the basis of one type of invisible ink.

Air. Oxygen. Ozone

AIR

For about two thousand years after the time of Aristotle air was believed to be an element. About 1660 A.D. a group of English scientists, all founder members of the Royal Society, performed experiments that cast much doubt on this belief. They were Mayow, Hooke, and Boyle. These experimenters showed that if certain processes were carried out in an enclosed space containing air over water (the famous 'bell-jar' experiments) in each case only one-fifth of the air was used up (shown by water rising in the jar). The processes were (a) burning (illustrated with phosphorus); (b) breathing (mouse enclosed in the bell jar); (c) rusting of damp iron (a much slower process). Mayow, Hooke and Boyle, became convinced that one-fifth of the air was different from the rest and at various times this fifth became known as 'fire air', 'vital air', 'active air', 'dephlogisticated air' (Priestley) and finally 'oxygen', by Lavoisier. In the remaining four-fifths of the air nothing would burn, no animal could live, and iron would not rust. It was known as 'inactive air' and by Priestley as 'phlogisticated air' (Section 79). Mayow also discovered that nitre (potassium nitrate) showed a kinship with the active part of the air. He found that gunpowder could continue to burn under water—nitre supplying for the combustion that which normally came from the air. In 1774, Priestley discovered a new gas by heating red calx (oxide) of mercury with the Sun's rays concentrated by a large lens. The calx was enclosed in a bell jar over water. He found that things burnt in the new gas more vigorously and for longer than in ordinary air and that a mouse could live in it for a longer time. About the same time, and independently, Scheele in Sweden obtained the new gas from nitric acid, nitre, red calx of mercury, and red lead (the oxide of lead, Pb_3O_4 (Section 44)).

Although it was suspected that the active fifth of the air and the new gas were one and the same it was, at that time, a very difficult thing to prove. Soon after Priestley had discovered the gas he went to Paris and told Lavoisier about

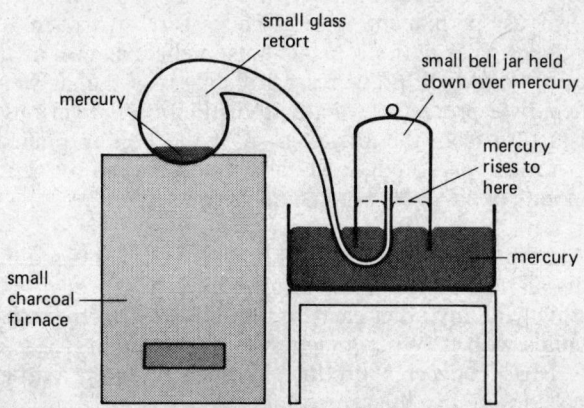

Figure 9 Lavoisier's method for the extraction of oxygen from air

it and it was Lavoisier who solved the problem. He heated a little mercury, contained in a small glass retort, for some days over a charcoal furnace. The end of the retort was bent round so as to emerge in a small jar held over mercury, the levels inside and out being made the same by sucking air out of the jar with the help of a piece of bent glass-tubing (Figure 9). Over a long period of time, a black powder gradually formed on the surface of the mercury in the retort (which became red when cold) and the level of the mercury in the jar rose slowly as part of the air was used. Lavoisier believed that the mercury was combining with the active fifth of the air to form the calx. Subsequently, the heating was discontinued and, in a remarkable quantitative experiment, Lavoisier recovered the red calx and, heating it to a higher temperature, obtained from it exactly the same quantity of pure 'oxygen' that had been abstracted from ordinary air during his first experiment. The volume of the gas was only about $130 \, cm^3$. This fundamental experiment made use of the reversible reaction between mercury and oxygen that would now be expressed:

$$2Hg(l) + O_2(g) \rightleftharpoons 2HgO(s)$$

The inactive four-fifths of the air was named by Lavoisier 'azote' meaning 'lifeless', a very good name for a comparatively inert gas. In this country we use the name 'nitrogen' meaning 'source of nitre', which is not so appropriate.

Oxygen and nitrogen together make up about 99% by volume of ordinary dry air. At the end of the nineteenth century, the noble gases were discovered and the other 1% of the air was found to be mainly argon. Carbon dioxide is also present in small quantity (normally about 0.03%). Some water vapour is always present in the atmosphere but, as its amount is so variable, it is customary to give the volume composition of *dry* air:

Nitrogen		78%
Oxygen		21%
Argon	nearly	1%
Carbon dioxide		0.03%

Very small quantities of other noble gases – neon, krypton, xenon.

Faraday and others succeeded in liquefying many common gases using low temperatures and high pressures, but three gases resisted their efforts and became known as 'permanent' gases. They were hydrogen, oxygen, and nitrogen. By making use of the Joule-Thomson effect as well as the necessary low temperatures and high pressures, all the gases of the air were eventually liquefied. Then, by a process of fractional distillation the various gases could be collected in turn from the air. Even so, if the boiling points were close, complete separation was still difficult – as with argon and oxygen. The respective boiling points of nitrogen, argon, and oxygen are: $-196 \, ^\circ C$, $-186 \, ^\circ C$, and $-183 \, ^\circ C$.

Nitrogen in the air acts almost entirely as a diluent and the chemical activity of the air is mainly that of diluted oxygen. Nitrogen itself does not combine readily with any other element, but the very electropositive ones will form compounds with it. When magnesium burns in air, for example, magnesium oxide, MgO, is the chief product, but there is always a little magnesium nitride, Mg_3N_2, formed at the same time.

Fairly pure nitrogen can quite easily be obtained from the air in the

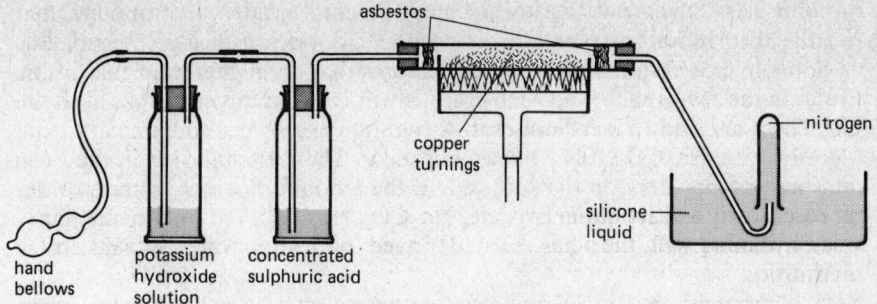

Figure 10 The laboratory preparation of nitrogen from air

laboratory (Figure 10). With the help of hand-bellows, puffs of air can be pushed intermittently through an apparatus containing three absorbents: first, potassium hydroxide solution to take out the small quantity of carbon dioxide; second, concentrated sulphuric acid to remove water vapour and spray from the alkali solution so that the red-hot tube that follows is not cracked; third, a hard-glass tube containing copper turnings, raised to red-heat, to remove the oxygen by the formation of black copper(II) oxide, CuO, superficially.

Nitrogen cannot be collected by the displacement of air because the relative densities of the gases are too close 14:14.4. If required dry, the nitrogen has to be collected over an appropriate silicone liquid (Section 61). The nitrogen, which will still contain about 1% of the noble gases when collected, will be found to extinguish a burning splint, but burning magnesium ribbon will continue to burn in the gas – though with much less vigour.

Water in the atmosphere

There is always a certain quantity of water vapour in the atmosphere and at any particular temperature there is a limit to the amount that the air can hold. This is known as the **saturation vapour pressure (svp).** At 14 °C it is only 12 mm. This small amount of water has strange effects on some elements and compounds. Sodium, for example, is slowly attacked by damp air and eventually becomes a solution of sodium hydroxide which, in turn, reacts with the carbon dioxide in the air. Many compounds attract a little water and become damp. These are said to be **hygroscopic.** (The prefixes 'hydro' (Latin) and 'hygro' (Greek) always refer to water.) Hygroscopic substances do not form saturated solutions and often the effect is due to the formation of hydrates (as with concentrated sulphuric acid) or other combination with water (as with calcium oxide).

A few compounds absorb so much water from the air that they dissolve in it and form solutions. If the vapour pressure of such a solution is lower than that of the water vapour in the atmosphere, as soon as a drop of saturated solution forms on the compound, the process will continue until the solid is dissolved. Such compounds are said to be **deliquescent** (from the Latin meaning 'to become liquid'). Deliquescent compounds include sodium hydroxide, iron(III) chloride, and chlorides and nitrates of magnesium and calcium. Compounds that deliquesce can only be kept pure in a desiccator – an apparatus for keeping things dry. The drying chamber of a desiccator has a greased ground-glass cover and the lower compartment contains a compound that readily absorbs water, e.g. anhydrous calcium chloride or silica gel (Section 61).

Some hydrated salts have vapour pressures that are higher than that of the water in the atmosphere and such crystals will *lose* water when exposed to the air. These are said to be **efflorescent.** A common example is sodium carbonate decahydrate, $Na_2CO_3,10H_2O$ (washing soda). These crystals, left in the open air, become powdery on the surface and the amount of water in the powder is much less; it is the monohydrate, Na_2CO_3,H_2O.

A hydrated salt that has been deprived of all its water is said to be **anhydrous.**

A compound that is deliquescent in this country, e.g. hydrated calcium chloride, $CaCl_2,6H_2O$, can be efflorescent in parts of Canada where the air is very dry and the water vapour pressure very low.

The air and respiration

When air is taken into the lungs, the oxygen forms a loose compound with the haemoglobin of the red blood corpuscles called oxyhaemoglobin. This blood, bright-red in colour, is pumped by the heart through the arteries to all parts of the body, here an exchange takes place – oxygen being given up and carbon dioxide being taken up. The blood, now dark-red in colour, returns to the heart via the veins. Arteries are, for protection, always deep-seated in the body, the veins being nearer the surface. About one-fifth of the oxygen inspired is taken up by the blood; the air expired, which formerly contained 0.03% carbon dioxide by volume, will now contain as much as 3.5% of this gas. In normal breathing only about one-seventh of the capacity of the lungs is used, though when taking violent exercise the rate and depth of breathing are greatly increased. Breathing rates vary enormously from individual to individual – anything from 7 to 25 breaths per minute. The small amount of carbon dioxide inspired regulates the depth of breathing and if the proportion of carbon dioxide is artificially increased, deeper breathing results. Anaesthetists make use of this fact. If the percentage greatly increases, as in a crowded badly-ventilated room, headaches are caused and, eventually, slow suffocation through lack of oxygen.

OXYGEN, O_2

Lavoisier gave to the active fifth of the air the name 'oxygen', which means 'source of acid', because he believed that all acids had oxygen in their composition. He was, of course, wrong about this (Section 33) but the name has not been changed. The gas is essential to almost all living creatures and it supports combustion better than any other gas. Very few elements will not burn in oxygen either at normal or increased pressure. The usual test for oxygen is the relighting of a glowing splint. There are two other gases that have this ability because they decompose at relatively low temperatures to give a mixture of gases that contains a higher proportion of oxygen than the one-fifth of ordinary air. The gases are dinitrogen oxide (nitrous oxide), N_2O, and dinitrogen tetroxide, N_2O_4, (Section 62). In elementary chemistry these are not likely to cause confusion. When ammonium nitrate is heated, dinitrogen oxide is formed but accompanied by steam and a glowing splint is not relit. Dinitrogen tetroxide is, when hot, dark-brown in colour and thus easily distinguished. Oxygen gas is readily absorbed by sodium pyrogallate solution (an organic acid – pyrogallic – with sodium hydroxide solution). During the absorption the solution becomes dark-brown in colour and heat is evolved.

Quite a large number of solid substances will evolve some oxygen on pyrolysis. They include all the common nitrates (except ammonium), several 'higher' oxides (Section 32) including peroxides, MnO_2, Pb_3O_4, PbO_2, etc., thermally unstable oxides such as HgO, chlorates, etc. Pure oxygen can be obtained by electrolysis of dilute sulphuric acid. Its usual laboratory and industrial preparations are given in Section 75.

OZONE, O_3

Ozone is an allotrope of oxygen (Section 50) and has three atoms in its molecule. The molecular shape is angular:

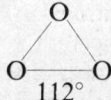

Ozone is a blue gas of b.p. $-112\,°C$. Ozonized oxygen can be prepared by passing a silent electric discharge continuously through oxygen gas. A concentration of up to 10% ozone can be obtained and it can be separated from oxygen by fractional liquefaction. It is, like hydrogen peroxide (Section 32), a powerful oxidizing agent, bleaching agent, and germicide. It is used in air purification systems. It is an endothermic substance (Section 46) and can be dangerously explosive when concentrated. If ozonized oxygen containing 10% by volume is heated, the ozone forms oxygen with the volume change 2 volumes of ozone giving 3 volumes of oxygen. In this case, therefore, if measurements are made under the same conditions, $100\,cm^3$ of ozonized oxygen containing 10% ozone would give $105\,cm^3$ oxygen after the decomposition of the ozone. Ozone forms addition compounds with unsaturated hydrocarbons (Section 54) and this is the reason why it is readily absorbed by turpentine. (Turpentine consists chiefly of an unsaturated hydrocarbon with the formula $C_{10}H_{16}$.) Ozone has a strong characteristic smell. The presence of ozone causes mercury to 'tail' through superficial oxidation of the metal. Ozone plays an important role in the upper atmosphere by absorbing short-wave ultra-violet radiation, thus screening the Earth's surface from these highly active rays. In our own atmosphere it can form as a pollutant (Section 74).

31 Oxidation and Reduction. Oxidation Number

OXIDATION AND REDUCTION

Oxidation is one of the processes in chemistry which started as a simple idea but has developed into a far broader concept. After the discovery of oxygen, oxidation was simply a process in which oxygen was added to another element or compound. Conversely, reduction involved the removal of oxygen. It was soon realized, however, that the removal of hydrogen from a compound often produced the effect of oxidation whereas the addition of hydrogen (often causing the removal of oxygen in the form of water) corresponded to reduction.

During Faraday's researches into the phenomena associated with electrolysis, he introduced the term 'ion' which simply means something that moves or goes – anions to the anode and cations to the cathode. Faraday could not, of course, define an ion in terms of atoms and electrons but, following his work, all acids, alkalis, and salts were regarded as having positive and negative parts. Hydrochloric acid gave hydrogen$(+)$ and chloride$(-)$ ions; potassium hydroxide gave potassium$(+)$ and hydroxide$(-)$ ions; sodium sulphate gave sodium$(+)$ and sulphate$(-)$ ions in the ratio 2:1. The term oxidation was widened to include any relative increase in the electronegative part of a compound. Not only was $PbO \longrightarrow PbO_2$ regarded as oxidation, but also $FeCl_2 \longrightarrow FeCl_3$. On the other hand, reduction became the relative increase in the electropositive part of a compound or a relative decrease in the electronegative part. Considering the two sulphides of copper, $Cu_2S \longrightarrow CuS$ is oxidation because, if one keeps the proportion of copper constant in each formula $Cu_2S \longrightarrow (Cu_2S_2)$, one can see that there has been a relative increase in the electronegative part.

With the discovery of the small particles that make up an atom, it was soon realized that oxidation can be defined in terms of gain or loss of electrons. *Loss of electrons* by an atom or group of atoms is **oxidation.** For example, $Cu^+ \longrightarrow Cu^{2+}$ is oxidation because the atom has lost one more electron. $Cl_2 \longrightarrow 2Cl^-$ is reduction because the covalent chlorine molecule has been changed to two ions each with one extra electron. *Gain of electrons* by an atom or group of atoms is **reduction.** $Fe^{3+} \longrightarrow Fe^{2+}$ is reduction because the iron ion has gained an electron. During electrolysis, *anodic processes involve oxidation* because it is at the anode that electrons are given up. Conversely, *cathodic processes involve reduction.* It follows also that the elements at the top of the activity series (Section 6) are powerful reducing agents (e.g. Na, Mg, Al) because their atoms readily lose electrons. The atoms of elements at the bottom of the series accept electrons readily (e.g. Cl, O, F) and are powerful oxidizing agents.

The reactions referred to in Section 6 as 'replacement' reactions can now be seen as examples of oxidation/reduction. The reaction in which metallic zinc replaces copper from a solution of copper(II) sulphate:

$$Zn(s) + Cu^{2+}SO_4^{2-}(aq) \longrightarrow Zn^{2+}SO_4^{2-}(aq) + Cu(s)$$

can be written more simply as:

$$Zn(s) + Cu^{2+}(aq) \longrightarrow Zn^{2+}(aq) + Cu(s)$$

All that has happened in this reaction is the transfer of two electrons from the zinc atom to the copper atom. The zinc atoms have lost electrons to become ions and the copper(II) ions have gained electrons to form metallic copper. In other words, zinc has acted as a reducing agent in precipitating the copper. The sulphate ions are 'spectator' ions.

Similarly, when chlorine replaces bromine from the solution of a bromide or iodine from the solution of an iodide, chlorine acts as an oxidizing agent in taking electrons from the other halide ions to form chloride ions. Here, the metal ions are 'spectator ions:

$$Cl_2(g) + Mg^{2+}2Br^-(aq) \longrightarrow Br_2(l) + Mg^{2+}2Cl^-(aq)$$
$$Cl_2(g) + 2K^+2I^-(aq) \longrightarrow I_2(s) + 2K^+2Cl^-(aq)$$

Not all chemical reactions, however, involve electron transfer. To quote a simple example: copper(II) oxide can be reduced to copper by heating it in a stream of hydrogen gas. Hydrogen consists of covalent molecules and copper(II) oxide is a mainly covalent compound; so the reaction can properly be represented by the equation:

$$H_2(g) + CuO(s) \longrightarrow Cu(s) + H_2O(g)$$

In the above reaction there is no electron exchange. The hydrogen is oxidized to steam and the copper(II) oxide is reduced to the metal. As so often happens, oxidation and reduction go on simultaneously – what one gains is lost by the other. Reactions of this character are known as **'redox'** reactions. There are cases known, however, where two oxidizing agents lose oxygen at the same time as, for example, the reaction of potassium manganate(VII) solution with hydrogen peroxide where both are mutually reduced with evolution of oxygen from each.

Any element is oxidized when it combines with another element that is more electronegative than itself. When an element combines with chlorine, it is oxidized as definitely as if it combines with oxygen. When oxygen fluoride, OF_2, is formed, even oxygen is oxidized! (It is better to regard this compound as oxygen fluoride rather than fluorine oxide because the more electronegative element should be put last.)

To sum up, processes that involve *oxidation* are

(1) *The combination of an element with an element more electronegative than itself.* Examples:

Burning of magnesium in oxygen

$$2Mg + O_2 \longrightarrow 2MgO$$

Combination of iron with sulphur to give iron(II) sulphide

$$Fe + S \longrightarrow FeS$$

(2) *The removal of hydrogen from a compound.* Examples:
An oxidizing agent liberating chlorine from hydrochloric acid

$$2HCl + (O) \longrightarrow Cl_2 + H_2O$$

Formation of ethanal by oxidation of ethanol (Section 56)

$$CH_3.CH_2OH + (O) \longrightarrow CH_3CHO + H_2O$$

(3) *Loss of one or more electrons from an atom, group of atoms, or ion.* Examples:
Sodium burning in chlorine gas

$$2Na + Cl_2 \longrightarrow 2Na^+2Cl^-$$

Replacement reactions described earlier in this section.

Conversely, *reduction* involves the loss of oxygen, the gain of hydrogen, or the gain of electrons. Examples:
(1) *Oxygen loss*
Copper(II) oxide reduced by heating in hydrogen gas

$$CuO + H_2 \longrightarrow Cu + H_2O$$

Lead(II) oxide reduced by heating with carbon

$$PbO + C \longrightarrow Pb + CO$$

(2) *Hydrogen gain*
Direct combination of sulphur vapour with hydrogen

$$H_2 + S \longrightarrow H_2S$$

Ethene converted to ethane by hydrogenation (Section 54)

$$C_2H_4 + H_2 \longrightarrow C_2H_6$$

(3) *Electron gain*
Reduction of copper ions in the replacement reactions described earlier in this section
Reduction of chlorine gas in the replacement reactions described earlier in this section.

OXIDATION NUMBER

Oxidation number helps to determine when oxidation or reduction has taken place, particularly in difficult cases. It is extremely useful for understanding the action of nitric acid on various metals (Section 64). The oxidation number of any common element is its normal valence figure, but putting a plus sign to that of an electropositive element and a minus sign to that of an electronegative one. Examples:

$+3$ aluminium
$+2$ calcium, magnesium, zinc
$+1$ potassium, sodium
-1 chlorine, bromine, iodine
-2 oxygen

Hydrogen is normally $+1$, but when combined with a very electropositive element it can become -1 (as in sodium hydride, NaH).

In the formula of any stable compound, if all the oxidation numbers of the elements involved are added together then the total must be zero.

For example, the oxidation number of sulphur in sulphuric acid, H_2SO_4, must be $+6$ because the two hydrogen atoms give $2 \times +1$ and the four oxygen atoms give 4×-2 so, to make the total add to zero, the sulphur atom must be $+6$. This gives another useful definition of oxidation:

Oxidation has taken place when the oxidation number of the element concerned has increased in a positive direction, e.g. $-3 \longrightarrow -2$; $-1 \longrightarrow +1$; $+6 \longrightarrow +7$; and so on.

Examples:
(1) *Oxides of lead:* PbO_2 changing to Pb_3O_4; relative to the lead there has been oxygen loss. This can be seen if the lead is made Pb_3 in each case: $(Pb_3O_6) \longrightarrow Pb_3O_4$. Using the oxidation number method it is clear that the process is one of reduction. As each oxygen is -2, the oxidation number of lead in the first of the two oxides is $+4$. In the second oxide the oxidation number is $+2\frac{2}{3}$. (There is no objection to fractional values for oxidation numbers.) As the oxidation number has been *decreased* by the change it must be an example of reduction.

(2) *Potassium manganate*(VII), $KMnO_4$: When potassium manganate(VII) is made on an industrial scale from manganese(IV) oxide, MnO_2, potassium manganate(VI), K_2MnO_4, which is green in colour, is formed first and then the purple manganate(VII) is obtained from this. Is the process $K_2MnO_4 \longrightarrow KMnO_4$ an example of oxidation or reduction? In the first compound the oxidation number of the manganese is $+6$; in the second it is $+7$. Therefore oxidation must have taken place and, on an industrial scale, chlorine is used as the oxidising agent.

(3) *Potassium dichromate*, $K_2Cr_2O_7$: Both potassium manganate(VII) and potassium dichromate are used in the laboratory as tests for sulphur dioxide (Section 68). Both manganese and chromium are transition elements (Section 70) and chromium can have valencies ranging from 2 to 6. A solution of potassium dichromate is orange in colour (due to the dichromate ions) but when hydroxide ions are added the colour becomes yellow, due to the formation of chromate ions. It is a reversible reaction because the addition of hydrogen ions (from an acid) reverses the colour to orange again. Written

in ionic form the reactions can be represented:

$$Cr_2O_7{}^{2-} + 2OH^- \longrightarrow 2CrO_4{}^{2-} + H_2O$$

orange   dilute   yellow
dichromate   alkali   chromate

$$2CrO_4{}^{2-} + 2H^+ \longrightarrow Cr_2O_7{}^{2-} + H_2O$$

yellow   dilute   orange
chromate   acid   dichromate

Is the change from dichromate to chromate oxidation or reduction? Working out the oxidation numbers shows that it is neither. The oxidation number is found to be the same in each case, $+6$.

One disadvantage of the oxidation number method is that it is much more difficult to apply where there is any **catenation** of atoms. An element may have atoms that have the ability to form chains when they link covalently to one another. Whenever this happens the oxidation number is of little help. As carbon (Group IV), more than any other element, can exhibit catenation by the linkage of carbon atoms to form chains (Section 52), the concept can rarely be applied in the field of organic chemistry. Other elements can show catenation to a lesser degree, particularly silicon (Group IV), nitrogen, phosphorus (Group V), oxygen, sulphur (Group VI), and these are all non-metals. Even in a simple compound like hydrogen peroxide, H_2O_2 (Section 32), the oxidation number method runs into difficulty. In water, H_2O, hydrogen is $+1$ and oxygen -2. What then of H_2O_2? Hydrogen peroxide, like water, is a covalent compound and its molecule has a definite shape (Section 32). The oxygen atoms are known to be linked to each other in the molecule and this can be indicated simply as H–O–O–H. As there is catenation the oxidation number method is not helpful.

32   Oxides. Hydrogen Peroxide

OXIDES

An oxide is a binary compound of oxygen with another element. Referring to the activity series given in Section 6, oxides are known of every element in the list but oxides of the least electropositive metals are very unstable. So also, as one might expect, are the oxides of the halogens.

Oxides are classified into six main groups as follows:

Basic oxides are chiefly those of the metals where the valence of the element concerned is the common one. Being bases (Section 35) each will combine with an acid to give a salt and water *only*. Those that react with water form alkalis but only the oxides of very electropositive elements – those of Group I and some of Group II – do this.

Acidic oxides are formed by non-metals (electronegative elements) and behave in the opposite way to basic oxides. Many react with water to give acids (Section 33) and these are known as acid anhydrides. (An anhydride (from the Greek meaning 'without water') is a residue that is formed when water is removed from a compound, the compound being reformed when water is united with it.) An acidic oxide will react with an alkali to form a salt and water *only*. Silicon oxide (sand), SiO_2, has no reaction with water but it will react with sodium hydroxide to give sodium silicate and water. Examples: SiO_2; P_2O_3; P_2O_5; I_2O_5; CO_2; SO_2; SO_3; N_2O_4.

Amphoteric oxides can be either basic or acidic under the appropriate conditions. They will react with caustic alkalis to give one type of salt and with acids to give another. Aluminium oxide, for example, can give aluminium salts with acids but aluminates with alkalis. Examples: Al_2O_3; As_2O_3 (arsenic(III) oxide); ZnO.

Neutral oxides have neither basic nor acidic properties and belong to no other class. Examples: H_2O; CO; N_2O.

Compound oxides have the chemical properties associated with two oxides of a metal but in combination. The two common examples in elementary chemistry are Fe_3O_4 and Pb_3O_4. The black oxide of iron, Fe_3O_4, is essentially basic because it behaves, in its chemical reactions, as if it were iron(II) oxide with iron(III) oxide: $FeO.Fe_2O_3$. It reacts with hydrochloric acid to give a mixture of iron(II) and iron(III) chlorides. It is not an oxidizing agent. 'Red lead', Pb_3O_4, has the properties associated with two units of lead(II) oxide with one unit of lead(IV) oxide $(PbO)_2.PbO_2$. Lead(IV) oxide has oxidizing power and so, therefore, has red lead.

Peroxides are salts of hydrogen peroxide and have the typical peroxide structure (see below). Examples: Na_2O_2; BaO_2.

Terms sometimes used such as 'higher oxide' or 'suboxide' merely indicate that the element has more or less oxygen in its oxide than its normal valence would warrant.

Although Lavoisier was wrong in assuming that all acids must contain oxygen in their composition (Section 33), oxygen and acidity are closely connected. Several elements form more than one compound with oxygen and *always* the higher the proportion of oxygen in the compound, the more acidic is its nature. Water, H_2O, for example, is neutral but hydrogen peroxide, H_2O_2, is a weak acid. The transition element manganese (Section 70) has as many as five different oxides and as the proportion of oxygen increases, so also does the acidity. Manganese(II) oxide, MnO, is basic; manganese(IV) oxide, MnO_2, is amphoteric. The oxide that contains the highest proportion of oxygen, manganese(VII) oxide, Mn_2O_7 is definitely acidic, forming manganate(VII) salts.

HYDROGEN PEROXIDE

This is a higher oxide of hydrogen but, in this compound, the valence of hydrogen remains 1 and that of oxygen 2. It illustrates in a simple way the

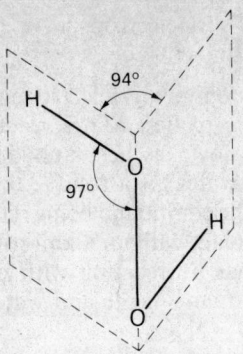

Figure 11   Hydrogen peroxide molecule

phenomenon of **catenation** where an atom of an element, in this case oxygen, is joined covalently to another atom of the same element. Hydrogen peroxide is mainly covalent and the shape of its molecule is well known. If one imagines a hinge opened at an angle of 94°, then the oxygen atoms would lie along the hinge-pin and the hydrogen atoms on each face of the hinge at at an angle of 97° with the oxygen line (Figure 11).

Hydrogen peroxide is slightly ionized and acts as a weak acid. Its salts, the peroxides, have structures similar to that of hydrogen peroxide itself, the two oxygen atoms being still joined to one another. Two well-known salts (both ionic) are sodium peroxide with the formula $Na^+ \ ^-O–O^- Na^+$ and barium peroxide, BaO_2,

$$Ba^{2+}$$

$$^-O–O^-$$

In this compound barium is divalent as in all its compounds.

Hydrogen peroxide is made on a large scale for use as an oxidant for rocket fuel. It is capable of providing a large volume of oxygen from a relatively small mass of peroxide:

$$2H_2O_2(l) \longrightarrow 2H_2O(l) + O_2(g)$$

Therefore 68 g peroxide can give one mole (22.4 litres at s.t.p.) of oxygen.

Pure hydrogen peroxide is difficult to obtain and that used as a rocket fuel is known as 'high-test' peroxide – over 95% pure. It is a colourless liquid and can be dangerously explosive if it contains oxidizable impurities. Various substances can catalyse its explosive decomposition. In everyday life, or in the laboratory, it is met with as '10-volume' or '20-volume' solutions in water and these are quite safe. Such solutions are commonly used as bleaching agents, antiseptics, etc. Hydrogen peroxide is a powerful oxidizing agent and therefore also a powerful germicide. Aqueous solutions of hydrogen peroxide gradually lose oxygen when stored, but manufacturers prevent this by incorporating one or more negative catalysts to retard decomposition.

The '20-volume' solution is such that 1 cm³ will provide (when catalytically decomposed with manganese(IV) oxide, MnO_2, for example) 20 cm³ of oxygen at s.t.p. The molarity of such a solution can readily be calculated. From the above equation, 22 400 cm³ of oxygen at s.t.p. can be obtained from 68 g of hydrogen peroxide. A '20-volume' solution gives 20 000 cm³ oxygen from 1 litre.

Therefore, $20\,000\,\text{cm}^3$ of oxygen must come from $\dfrac{68 \times 20\,000}{22\,400} = 60.7\,\text{g of } H_2O_2$

As the mole of hydrogen peroxide is $34\,\text{g}$ the molarity of a '20-volume' solution must be $\dfrac{60.7}{34} = 1.8$ (approx.)

Thus a '10-volume' solution must be a little less than molar (0.9).

Methods of obtaining hydrogen peroxide

In the laboratory, a dilute solution can be obtained by adding barium peroxide to moderately concentrated sulphuric or orthophosphoric acid. (Barium peroxide itself is prepared by heating barium oxide, BaO, in air or oxygen at 500 °C.) In the cases of the acids mentioned, each produces an insoluble salt which is convenient for the separation of the products. Hydrogen ions catalyse the decomposition of hydrogen peroxide, so the acid must not be too concentrated or present in any great excess. Heat also decomposes hydrogen peroxide, so the operation is carried out surrounded by a freezing mixture. The insoluble salt is removed by filtration and the filtrate is found to give the characteristic tests for hydrogen peroxide:

$$BaO_2(s) + H_2SO_4(aq) \longrightarrow BaSO_4(s) + H_2O_2(aq)$$

Such methods of ion aggregation (Section 18) are too inefficient for the industrial preparation of hydrogen peroxide and other, very different, methods have been devised. They include:

(1) Direct synthesis from the elements under special conditions.
(2) Electrolysis of ammonium sulphate solution with excess sulphuric acid using platinum electrodes and high current density.

Tests for hydrogen peroxide

As a powerful oxidizing agent, hydrogen peroxide gives all the usual tests for such agents. These include:

(1) Liberation of chlorine from hydrochloric acid.
(2) Liberation of iodine from an acidified solution of potassium iodide.
(3) Oxidation of Fe^{2+} to Fe^{3+}.
(4) Oxidation of black lead(II) sulphide to white lead(II) sulphate:

$$PbS + 4(O) \longrightarrow PbSO_4$$

As nearly all oxidizing agents give the first three reactions listed above, it is helpful to have one or two specific tests for hydrogen peroxide, i.e. tests which give results for this compound alone. Here is one.

(5) If a hydrogen peroxide solution is added to a mixture of orange-coloured potassium dichromate solution with dilute sulphuric acid, a blue colour is seen which quickly fades when the mixture is shaken. To obtain a lasting effect, it is usual to put a small layer of ethoxyethane (ether) on the top of the

mixed solutions before adding the suspected hydrogen peroxide solution. The blue colour is probably due to the formation of a higher oxide of chromium, perhaps CrO_5, and if the ethoxyethane is present the blue compound enters the ethoxyethane layer where the colour will remain. It is an extremely sensitive test and will detect hydrogen peroxide in minute concentration.

33   Acids

For at least a thousand years, acids have been looked upon as a special class of compounds because they have simple properties in common: sour taste (safely diluted if necessary); ability to change the colour of certain dyes (e.g. litmus – blue to red); ability to neutralize alkalis; power to dissolve chalk; corrosive action on common metals.

Lavoisier had wrongly concluded that all acids must be derived from oxygen (Section 30). If he had been allowed to live the normal span of life (he was guillotined in 1794 during the French revolution at the age of 50) he would have known of Davy's isolation of the alkali metals which burn in oxygen to give products that, when shaken with water, produce **alkalis**. Furthermore, acids were prepared that had no oxygen in their composition and chemists finally decided that the element common to all acids is **hydrogen**.

During the twentieth century, it was discovered that the liquid in which a compound is dissolved may play an important part in producing an acid. Previously chemists had found that *dry* hydrogen chloride has no acid properties. As soon as it is dissolved in water, a strong acid is obtained. This is now simply explained. Hydrogen chloride is a covalent compound consisting of separate molecules with the formula HCl. When the gas is passed into water there is a chemical reaction (energy being released) and hydroxonium (proton H^+ attached to a water molecule) and chloride ions are formed thus:

$$HCl(g) + H_2O(l) \longrightarrow H_3O^+(aq) + Cl^-(aq)$$

Hydrogen chloride gas will dissolve quite well in various organic solvents but in no case is an acid solution formed. Chemists have therefore tried to find a definition of an acid that is independent of the solvent.

In elementary chemistry, water is the only solvent in which acids are likely to appear, so a useful and adequate definition of an acid is as follows.

An acid is a compound that gives, in aqueous solution, hydroxonium as the only positive ions.

The word 'only' distinguishes an acid from an acid salt (Section 37). A more general definition, quite independent of the solvent, is the one proposed independently in 1923 by Brønsted in Denmark and Lowry in England:

An acid is a compound that has a tendency to lose a proton.

Basicity

Simple acids, such as hydrochloric and nitric, lose only one proton per 'molecule' and are known as **monobasic** acids:

$$HCl \longrightarrow H^+ + Cl^-$$
$$HNO_3 \longrightarrow H^+ + NO_3^-$$

Sulphuric acid, H_2SO_4, can ionize in two stages and the 'molecule' can provide two protons. It is therefore known as a **dibasic** acid:

$$H_2SO_4 \longrightarrow H^+ + HSO_4^- \text{ (hydrogensulphate ('bisulphate') ion)}$$
$$HSO_4^- \longrightarrow H^+ + SO_4^{2-} \text{ (sulphate ion)}$$

Orthophosphoric acid, H_3PO_4 (Section 66), is a **tribasic** acid, but most common acids are either monobasic or dibasic.

Ethanoic (acetic) acid, $CH_3.COOH$ (Section 56), has four atoms of hydrogen in its molecule but is only able to lose one proton (the one in the carboxyl group) and is a monobasic acid:

$$CH_3.COOH \longrightarrow H^+ + CH_3COO^- \text{ ethanoate (acetate) ion}$$

Strong and weak acids

If an acid in aqueous solution consists entirely of ions it will be a strong acid. It is contributing to the solution as many protons as it possibly can and is said to be completely **dissociated** (or **ionized**). This is approximately true for the three most common mineral (inorganic) acids: hydrochloric, HCl; nitric, HNO_3; sulphuric, H_2SO_4. Ethanoic acid (Section 56) is, on the other hand, a weak acid because it is only slightly dissociated when concentrated and relatively few of its molecules give protons in aqueous solution. The strength of an acid *cannot* be judged by its corrosiveness or by its power to drive another acid from its salts. Hydrochloric has not the corrosive properties of nitric acid, but it is just as strong an acid. Sulphuric acid can drive hydrogen chloride from the chloride of a metal or nitric acid from a nitrate only because sulphuric acid is non-volatile (does not form a vapour easily) whereas the other two compounds are volatile (Section 48).

'Strength' must not be confused with 'concentration'. The concentration of an acid is simply the mass of the acid – strong or weak – that is dissolved in a definite volume of solution (usually 1 litre). A molar solution (M.) of an acid contains one mole (Section 3) of the acid per litre of solution at a stated temperature (usually $20\,°C$). M.HCl will contain $36.5\,g\,l^{-1}$; $M.HNO_3$, $63\,g\,l^{-1}$; $M.H_2SO_4$, $98\,g\,l^{-1}$; $M.CH_3.COOH$, $60\,g\,l^{-1}$. These all have equal molar concentrations but whereas the first three are strong acids, the last one is weak. Dilute hydrochloric acid is still a strong acid because it is completely ionized; but ethanoic acid can never be a strong acid even when it is concentrated. In fact, ethanoic acid is more ionized in dilute solution but it never approaches complete dissociation.

The molar concentrations of the three common concentrated acids usually found in the laboratory are: hydrochloric, 12M; nitric, 15M; sulphuric, 18M. The dilute acids provided may be 1M or 2M.

34   Action of Acids with Metals

In the activity series (Section 6) hydrogen is found five places above the last metal. This means that the metals that come above hydrogen in the list should be able to replace hydrogen from acids, but with less vigour as the metal concerned becomes closer to hydrogen in the series. Thus magnesium ribbon disappears rapidly when placed in dilute hydrochloric acid whereas with tin there is little action unless the metal is boiled with concentrated hydrochloric acid. Metals that come below hydrogen in the series, e.g. copper and the noble metals, will not be affected by the acid in any way.

When magnesium reacts with hydrochloric acid, the magnesium metal loses electrons to become ions, whilst the hydrogen ions of the acid gain electrons to become atoms which then unite to form molecules of hydrogen gas:

$$Mg + 2H^+(2Cl^-) \longrightarrow Mg^{2+}(2Cl^-) + H_2$$

The chloride ions here are 'spectator' ions. From the resulting solution, by suitable evaporation, hydrated magnesium chloride, $MgCl_2, 6H_2O$, can be obtained in crystalline form.

Hydrogen chloride will react with heated magnesium to form anhydrous magnesium chloride and hydrogen, a molecular equation being appropriate:

$$Mg(s) + 2HCl(g) \longrightarrow MgCl_2(s) + H_2(g)$$

The action of concentrated hydrochloric acid with tin produces, in a similar way, hydrogen and ions of tin(II):

$$Sn + 2H^+ \longrightarrow Sn^{2+} + H_2$$

Tin(II) chloride crystals, $SnCl_2, 2H_2O$, can be obtained from the resulting solution. The anhydrous salt, as in the case of magnesium, can be obtained by passing hydrogen chloride gas over heated tin.

With *dilute* sulphuric acid the action is much the same as with hydrochloric acid. When zinc, for example, reacts with the dilute acid the metal loses electrons and zinc sulphate can be obtained from the resulting solution:

$$Zn + 2H^+(SO_4{}^{2-}) \longrightarrow Zn^{2+}(SO_4{}^{2-}) + H_2$$

Concentrated sulphuric acid, when heated with a metal, has a very different reaction. The acid itself is reduced to sulphur dioxide and, in some cases, even to hydrogen sulphide. *No* hydrogen is obtained but water instead. The equation usually written for the reaction with zinc is:

$$Zn + 2H_2SO_4 \longrightarrow Zn^{2+}SO_4{}^{2-} + SO_2 + 2H_2O$$

This is a balanced equation but it by no means represents all that takes place. A similar equation can be written for copper but, in this case, copper(II) sulphide is formed as well as the sulphate.

Hot concentrated sulphuric acid oxidizes all metals except the very noble ones such as gold and platinum. Unless water is present, the cold concentrated acid has only a very slow reaction.

The action of nitric acid with metals is much more complex because not only is it a strong acid but also a strong oxidizing agent. The products vary with the concentration of the acid, the reducing power of the metal, the temperature, and other factors (Section 64). It is impossible to write a single equation representing all that may go on. Hydrogen gas is hardly ever obtained. Magnesium with very dilute nitric acid can produce hydrogen gas but, in this case, the oxidizing power of nitric acid is very weak. Nitric acid attacks nearly all metals, including silver readily, but not gold or platinum. These noble metals are attacked by 'aqua regia' (Section 73).

NOTE: An acid will have very little effect on a metal if the salt to be formed is insoluble, because the metal will become coated with a layer of the insoluble salt and reaction will cease. The effect can be noticed when lead is treated with hot moderately-concentrated sulphuric acid.

35 Bases. Neutralization

BASES

A base is defined as any compound that will combine with an acid to form a salt and water only.

Most oxides and hydroxides of metals (in which the metal is exerting its normal valence) are bases.

Two simple examples of bases acting with acids:
Formation of copper(II) sulphate from copper(II) oxide

$$CuO + H_2SO_4 \longrightarrow CuSO_4 + H_2O$$

Formation of calcium chloride from calcium hydroxide

$$Ca(OH)_2 + 2HCl \longrightarrow CaCl_2 + 2H_2O$$

Brønsted and Lowry defined an acid (Section 33) as a compound that has a tendency to lose a proton and therefore, conversely, their definition of a base is

a base is a compound that has a tendency to gain a proton.

Acids are *proton donors* whereas *bases* are *proton acceptors*. The ammonium ion, NH_4^+, has to be regarded as an acid because it can donate a proton. Ammonia is regarded as a base because its molecule can accept a proton:

$$\begin{array}{ccccc} \text{ACID} & \rightleftharpoons & \text{BASE} & + & \text{PROTON} \\ \left[\begin{matrix} \overset{\displaystyle H}{\underset{\displaystyle \ddot{H}}{H\!:\!\ddot{N}\!:\!H}} \end{matrix} \right]^+ & \rightleftharpoons & \overset{\displaystyle H}{\underset{\displaystyle \ddot{H}}{H\!:\!\ddot{N}\!:}} & + & H^+ \end{array}$$

Relationship between base and alkali

If the oxide of a metal reacts with water (as is the case for most of the elements in Groups I and II) *or if the hydroxide of a metal dissolves in water, then an alkali is formed.*

Many metal oxides do not react with water and most hydroxides are insoluble so that, whilst there are many bases, there are not many common alkalis. Only five or six are to be found in an elementary laboratory. Any aqueous alkaline solution contains an excess of hydroxide ions and has many properties that are opposite to those of an acid.

Simple ideas about alkalis have been expressed for hundreds of years and the word itself goes back to the Arabic alchemists who prepared one alkali by calcining (i.e. strongly heating) land plants to obtain the ash which, when extracted with water, gave an alkaline solution. (The word 'alkali' means 'calcined ashes'.) Wood ash may contain up to 30% of potassium carbonate and a solution of this compound gives hydroxide ions by slight hydrolysis (as explained later in this section). The alchemists regarded a solution of 'pot-ash' as a mild alkali and they knew how to convert the mild alkali into a caustic one (in this case potassium hydroxide) by heating with a suspension of slaked-lime (calcium hydroxide). ('Caustic' because of its unpleasant action on the skin.) This old procedure has become an important industrial process (Section 38).

All alkaline solutions have characteristic properties which are almost always opposite to those of acids: soapy feel; effect on certain dyes – reversing the colour that acids produce with them; ability to neutralize acids; no effect on chalk; little action with common metals (except zinc and aluminium).

Just as the properties of aqueous acid solutions depend upon their excess of hydrogen ions, the properties of aqueous alkaline solutions are due to their containing an excess of hydroxide ions. If the hydroxide of a metal is completely ionized, it will give a strongly alkaline solution e.g. Na^+OH^- and K^+OH^-. Sodium and potassium hydroxides are known therefore as strong (or caustic) alkalis. Calcium hydroxide (slaked-lime), although ionized, is not very soluble in water and produces only a weakly-alkaline solution (lime-water). Ammonium hydroxide, which is a special case not derived from a metal, is only slightly ionized into NH_4^+ and OH^- and is therefore a weak alkali (Section 63).

Sodium carbonate and potassium carbonate, which the alchemists regarded as mild alkalis, are salts of carbonic acid, but they can act as weak alkalis because of slight hydrolysis brought about by water as the solvent (Section 20):

$$2Na^+ + CO_3^{2-} + 2H.OH \longrightarrow H_2CO_3 + 2Na^+ + 2OH^-$$

The small amount of carbonic acid formed is only slightly ionized, whereas the sodium hydroxide is completely ionized. This leads to an excess of hydroxide ions in the solution and an alkaline reaction. Carbonates are *not* bases, because when they are used to neutralize acids; carbon dioxide is always produced besides the salt and water. For example:

$$Na_2CO_3 + 2HCl \longrightarrow 2NaCl + H_2O + CO_2$$

The reaction is accompanied with brisk effervescence, which would never occur when a base neutralizes an acid.

The only alkalis commonly encountered in an elementary laboratory are:

Strong alkalis (completely ionized):

Sodium hydroxide	Na^+OH^-	(Caustic soda)
Potassium hydroxide	K^+OH^-	(Caustic potash)
Barium hydroxide	$Ba^{2+}(OH^-)_2$	(Not so soluble in water. Solution known as 'baryta water'.)

Weak alkalis:

Calcium hydroxide	$Ca^{2+}(OH^-)_2$	(Not very soluble in water. Solution known as 'lime-water'.)
Ammonium hydroxide	$NH_4^+OH^-$	(Only slightly ionized).

NEUTRALIZATION

When a mole of hydrochloric acid is neutralized by a mole of sodium hydroxide to form a salt, all that is happening is the formation of a mole of practically unionized water from a mole each of hydrogen and hydroxide ions, and the heat change is a measure of that reaction (Section 46):

$$H^+Cl^- + Na^+OH^- \longrightarrow Na^+Cl^- + H_2O$$

or simply
$$H^+ + OH^- \longrightarrow H_2O$$

Hydrogen ions, donated by the acid, are accepted by the hydroxide ions. Exactly the same result would be achieved, for example, by the neutralization of nitric acid, HNO_3, with potassium hydroxide, KOH.

If the salt formed is to be obtained in a pure state, it is essential that there should be some method of showing when the right mass of acid in solution has been added, so that the resulting salt is not contaminated with excess of either acid or alkali. This is done by the addition of a suitable dye or **indicator**. There exist a large number of such indicators, that are one colour in alkaline solution and a different colour (or no colour) in acid solution. Thus if the acid is added to the alkali until the indicator is at the point where it changes from one colour to the other, this will show that the correct amount of acid has been added. For accurate work the process of **titration**, using burette and pipette, is employed. Three indicators that are commonly used in the laboratory are shown in Table 12.

Litmus is not satisfactory for accurate work, so the usual indicators for acid/alkali titrations are methyl orange (for strong acids) and phenolphthalein (for weak acids). The **end-point** (i.e. the point where the indicator just begins to

Table 12

Indicator	Alkaline colour	Acid colour
Litmus	blue	red
Methyl orange	yellow	pink
Phenolphthalein	purple	colourless

change colour) is not easy to see when methyl orange is used, particularly in the yellowish light coming from an ordinary electric lamp, so many people prefer to use *screened* methyl orange. This consists simply of methyl orange solution mixed with a suitable blue dye. The colour change will thus be from green (yellow + blue) to purple (pink + blue) with an intermediate shade of grey just at the end-point and this change is certainly easier to see.

36   pH and the pH Scale

The acidity of an aqueous solution depends entirely on the concentration of the hydroxonium ions, H_3O^+, present. For simplicity, these are usually written as H^+ and hydroxonium concentration is shown as $[H^+]$, although H^+ is actually representing a proton attached to a water molecule when in aqueous solution.

Pure water is almost a non-conductor of electricity. It is mainly a covalent compound but it has a very slight dissociation into H_3O^+ and OH^-. It has been found that at 25 °C, if the hydroxonium concentration is multiplied by that of the hydroxide ion, the product is very nearly 1×10^{-14}. (The temperature must be stated as the result varies markedly with change of temperature.) This figure is known as the **ionic product for water.**

$$[H^+] \times [OH^-] = 1 \times 10^{-14} \text{ at } 25 \,^\circ\text{C}$$

As the number of hydrogen (hydroxonium) ions must be equal to the number of hydroxide ions, the concentration of the hydrogen ions in pure water must be 1×10^{-7} at 25 °C.

$$[H^+] = [OH^-] = 1 \times 10^{-7} \text{ at } 25 \,^\circ\text{C}$$

This means that in pure water the concentration of the hydrogen ions is only 0.000 000 1 M.

In any aqueous solution at a particular temperature, the ionic product for water remains constant; so that if a little acid is added to pure water putting up the hydrogen ion concentration to, say, 1×10^{-3} then the hydroxide ion concentration must drop correspondingly to 1×10^{-11}, thus keeping the product, as always, to 1×10^{-14}. One litre of this solution will contain 0.001 mole hydrogen and 0.00000000001 mole hydroxide ion.

To avoid the use of confusing indices and many noughts, Sørensen, in 1909, introduced the pH system for indicating acidity. 'p' stood for the 'puissance' (power) of the hydrogen ion concentration.

The 'pH' of any solution is the logarithm (to base 10) of the reciprocal of the hydrogen ion concentration or, more simply, the log of the hydrogen ion concentration with its sign changed.

As the hydrogen ion concentration in pure water is 1×10^{-7} (at 25 °C) this would be written as pH = 7, because the log of this number is −7.

The more acid the solution, the lower the pH figure will be. pH 1 is very acid, whereas pH 13 is very alkaline. To give a simple example let us calculate the pH of a 0.1M (decimolar) solution of hydrochloric acid. A molar solution of hydrochloric acid can be considered as completely dissociated, so a 0.1M solution will contain 0.1 mole of hydrogen ions. The hydrogen ion concentration of the decimolar acid will therefore be $1 \times 10^{-1} \, \text{mol} \, l^{-1}$. The log of this figure being −1, the pH will be 1.

Any solution is exactly neutral when its pH is 7.0 – the same value that pure water has. During a titration, the end-point (Section 35) is when the indicator changes colour but this by no means indicates that the solution is exactly neutral. Consider the addition of a weak acid, ethanoic acid to a strong alkali such as sodium hydroxide:

$$CH_3.COOH + NaOH \longrightarrow CH_3.COONa + H_2O$$

When the correct amount of acid has been added to a known volume of alkali, there will be no excess acid and no excess sodium hydroxide and the solution should contain only sodium ethanoate and water; but as sodium ethanoate is the salt of a weak acid it is slightly hydrolysed by water (Section 20) to give an excess of hydroxide ions and its pH is not 7.0 but somewhat above 8, i.e. on the alkaline side of exact neutrality. For this titration, therefore, we should need an indicator that changes colour at about pH 8. If, on the other hand, a weak alkali such as ammonia solution was being titrated against a strong acid such as nitric acid, the ammonium nitrate formed would also be somewhat hydrolysed to give, in a similar way, an excess of hydrogen ions. The equivalence point in this case would be on the acid side of neutral, say between pH 5 and pH 6 and we should need an indicator that would change colour in this area of pH. No indicator changes its colour sharply, but goes from one colour to the other over a range of about 1.5 units on the pH scale. By mixing indicators that do not interfere chemically with one another a wide-range indicator mixture can be produced that will change colour several times over a wide range of pH. Such indicators are known as 'universal' or simply 'wide-range' indicators.

If a simple pH scale is consulted, such as in Figure 12, it will show which indicator should be chosen for a particular titration of acid against alkali.

Below the scale are marked the ranges over which a few selected indicators change colour. If a strong acid is added slowly to a strong alkali, all the time that even a little alkali is in excess, the pH will be in the region 13–14. As soon as the right amount has been added, a little acid in excess will change the pH quite suddenly to the region 1–2. At the end-point, therefore, the pH of the

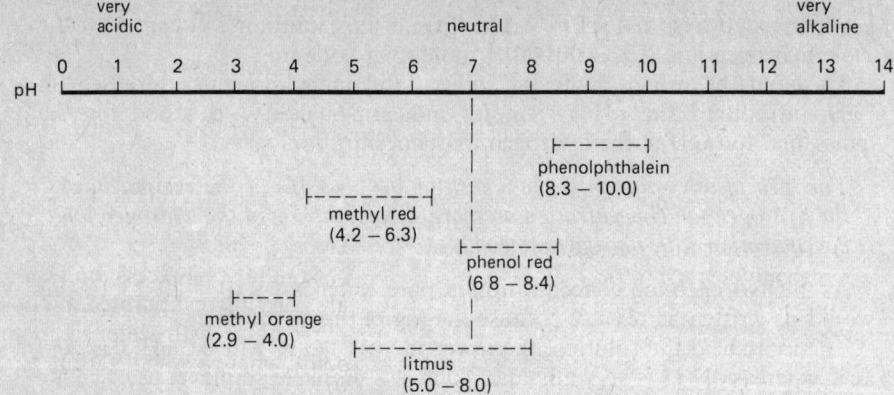

Figure 12　pH scale

solution runs quickly through the whole range of common indicators; so, for this titration, almost any indicator will do. For the titration of weak acid/strong alkali where, at the equivalence point, the pH will be 8–9, phenolphthalein would be a suitable choice. With a weak alkali/strong acid titration, the pH at the equivalence point may be 4–6 and methyl orange or methyl red would fulfil the requirement. The titration weak acid/weak alkali (e.g. ethanoic acid/ammonia) is never carried out, because it is not possible for any indicator to show when the correct amount of acid has been added. Various titrations and indicators are summarized in Table 13.

Table 13

Titration	Indicator
Strong acid/strong alkali	Any common indicator
Weak acid/strong alkali	Phenolphthalein
Strong acid/weak alkali	Methyl orange (or red)
Weak acid/weak alkali	No indicator suitable

37　Salts and their Preparation. Types of Salt

Whenever an acid and a base react together, a salt is formed and the salt contains one part derived from the base (cation – such as a metal or other positive ion, e.g. ammonium) and the other part from the acid (anion). If the metal is an element of Group I or II the derived salt is likely to be ionic.

Salts of metals of higher valence (such as aluminium(III) and lead(IV) are more likely to be covalent.

PREPARATION OF SALTS

The four chief methods for the preparation of salts used in the laboratory are:

(1) Whilst warming a dilute acid in a beaker, the appropriate solid (metal, base or carbonate) is stirred in until action ceases and the solid is in excess (no acid left). The excess solid is filtered off and the filtrate carefully evaporated, on a steam bath if available. If the salt is known to crystallize with water in the crystal lattice, the filtrate is evaporated to a point where crystallization is imminent (tested by removing a little of the solution on a glass rod) and then set aside to cool. When crystals have formed they are removed, rinsed quickly with water and dried between filter papers. If the salt is anhydrous, e.g. sodium chloride or lead nitrate, the filtrate can be evaporated very carefully to dryness. Examples:

(a) $ZnSO_4,7H_2O$ from granulated zinc and dilute sulphuric acid

Metal $\qquad Zn + H_2SO_4 \longrightarrow ZnSO_4 + H_2$

(b) Anhydrous $Pb(NO_3)_2$ from lead(II) hydroxide and dilute nitric acid

Base $\qquad Pb(OH)_2 + 2HNO_3 \longrightarrow Pb(NO_3)_2 + 2H_2O$

(c) Anhydrous $CaCl_2$ from calcium carbonate and dilute hydrochloric acid

Carbonate $\qquad CaCO_3 + 2HCl \longrightarrow CaCl_2 + H_2O + CO_2$

(2) Some simple salts and some acid salts can be made by titration, though this is a variant of (b) above, using a soluble base. When the correct relative volumes have been determined, the experiment is repeated mixing the correct quantities – this time without indicator – and the solution evaporated. Salts can be obtained from alkalis in this way.

(3) For insoluble salts, the method of ion aggregation (Section 18).

(4) Direct synthesis from elements. This method would only be used in special cases for *binary compounds* – particularly sulphides and halides. Examples:
(a) Iron filings heated with sulphur to give iron(II) sulphide.
(b) A metal heated in chlorine to give the chloride, e.g. anhydrous $MgCl_2$ or $ZnCl_2$. (Chlorine is such a powerful oxidizing agent that, if a metal has more than one valence, chlorine always forms the chloride of the higher valence, e.g. iron(III) chloride and tin(IV) chloride.)

A few salts could be prepared by only one of the methods outlined above, but with many others there could be a choice of two or three possible methods.

TYPES OF SALT

Acid salts

It was shown in Section 33 that some acids are mono-, di-, or tri-basic according to the number of hydrogen ions available from one 'molecule' of acid. Carbonic acid, H_2CO_3, is dibasic and from it can be derived a 'normal' carbonate containing the ion CO_3^{2-} and an 'acid' carbonate, usually now called 'hydrogencarbonate'(bicarbonate), containing the ion HCO_3^-. Although sodium hydrogencarbonate (sodium bicarbonate), $NaHCO_3$, is classed as an acid salt, it does *not* have an acid reaction in aqueous solution. Its pH is actually a little above 7. Sulphuric acid, H_2SO_4, is also dibasic forming normal sulphates (having SO_4^{2-} ions) and hydrogensulphates (bisulphates) (having HSO_4^- ions). In this case, sodium sulphate, Na_2SO_4, gives an aqueous solution that is neutral, whereas sodium hydrogensulphate is strongly acid in solution. It needs twice as much acid to form the hydrogensulphate as it does to form the normal sulphate from the same quantity of base:

$$2NaOH + H_2SO_4 \longrightarrow Na_2SO_4 + 2H_2O \qquad \text{(normal salt formed)}$$
$$(2)NaOH + (2)H_2SO_4 \longrightarrow (2)NaHSO_4 + (2)H_2O \quad \text{(acid salt formed)}$$

The two salts can readily be obtained by titration using the same volume of sodium hydroxide solution in each case, but *double* the volume of acid for the acid salt. Having obtained the two salts they can be distinguished from one another by heating with sodium chloride. The acid salt, sodium hydrogensulphate, still has some of the properties associated with sulphuric acid itself and will evolve hydrogen chloride when heated with the chloride:

$$NaHSO_4(s) + NaCl(s) \longrightarrow Na_2SO_4(s) + HCl(g)$$

The normal sulphate has no action when heated with the chloride.

Basic salts

This, as the name implies, covers salts that are partly salt and partly base (hydroxide or oxide). Sometimes the two are together in definite proportions as shown by the formulae, but quite often the composition is indefinite. Basic salts are formed when salt solutions are evaporated and a certain amount of hydrolysis occurs (Section 20). Neither anhydrous magnesium chloride nor anhydrous zinc chloride can be prepared by evaporating chloride solutions because the hot water hydrolyses the chloride to some extent giving off hydrogen chloride and leaving hydroxide ions in solution. If the gas escapes, the reaction continues left to right:

$$Cl^- + H.OH \longrightarrow OH^- + HCl(g)$$

(These chlorides can be obtained in anhydrous form by passing dry chlorine gas over the heated metal). If excess zinc reacts with dilute hydrochloric acid and the filtered solution is evaporated to dryness the basic salt $Zn(OH)Cl$ is obtained and not $ZnCl_2$.

Other examples are provided by the precipitation of some metal carbonates using sodium carbonate solution. If sodium carbonate solution be added to a lead(II) nitrate solution, a basic lead carbonate is precipitated the formula of which corresponds to $(PbCO_3)_2,Pb(OH)_2$. Because sodium carbonate is to some extent hydrolysed in aqueous solution, there are free hydroxide ions present which precipitate some of the lead as hydroxide whilst the carbonate ions precipitate most of the lead as carbonate. One way of avoiding the hydrolysis is by using sodium hydrogencarbonate solution as the precipitating agent. This compound gives almost entirely sodium and hydrogencarbonate ions with scarcely any hydroxide ions: $Na^+HCO_3^-$

With sodium carbonate:

$$Pb^{2+}(NO_3^-)_2 + (Na^+)_2CO_3^{2-} \longrightarrow PbCO_3\,(basic) + 2Na^+2NO_3^-$$

With sodium hydrogencarbonate:

$$Pb^{2+}(NO_3^-)_2 + 2Na^+2HCO_3^- \longrightarrow PbCO_3 + 2Na^+2NO_3^- + H_2O + CO_2$$

Even this method, however, is not always effective because the green copper(II) carbonate is always basic whichever carbonate is used to precipitate it from a copper(II) salt solution.

Complex salts

A complex salt contains a complex ion, i.e. a simple ion united to another ion or group so that it loses its identity and its normal properties. A complex ion is indicated by square brackets. Examples:

Iodine forming the $[I_3]^-$ ion (Section 23).
Lead(II) chloride dissolving in concentrated hydrochloric acid (Section 44).
Copper(II) tetrammine ion (Sections 29 and 72).

Double salts

These are produced when two simple salts crystallize together in definite proportions. They are quite common in nature, e.g. dolomite, a double carbonate, $CaCO_3,MgCO_3$; carnallite, a double chloride $KCl,MgCl_2,6H_2O$. The alums (Section 42) are all double salts.

38   The Alkali Industry

When chemists speak of the 'Alkali Industry' they refer chiefly to the manufacture of sodium carbonate and sodium hydroxide, both of which are needed by industry on a vast scale.

Sodium carbonate

The hydrated form of this compound, $Na_2CO_3,10H_2O$, is the familiar 'washing-soda'. The chief uses of the carbonate include paper-making, glass making, soap making, water softening, textile manufacture, dyestuff manufacture, in cleansing mixtures, and in the production of sodium hydroxide. The huge quantities produced each year are usually expressed in tonnes (a 'tonne' being a metric ton, i.e. 1000 kg). The method now used, the 'ammonia-soda' process was patented in 1838 but was not of great importance until it was developed by the Belgian chemist, Solvay, in 1872.

The raw materials for the ammonia-soda process are sodium chloride and limestone, with coal and coke as fuels; the final products are sodium carbonate and calcium chloride. The calcium chloride is produced in vast quantity, but, unfortunately, there is little use for it, and it is the only uneconomic product of the process. The overall reaction can be written:

$$2NaCl + CaCO_3 \longrightarrow Na_2CO_3 + CaCl_2$$

but there are many stages to be passed through to achieve this result.

In the first stage brine, pumped from salt mines, is saturated with ammonia gas in the 'ammonia scrubber' (Figure 13). Ammonia is passed into the base of a tower as the brine trickles down it. The ammonia-saturated brine then goes to the top of one or more Solvay (or 'carbonating') towers where it, in turn, passes downwards as carbon dioxide passes upwards (Figure 14). In the Solvay tower(s) ammonium hydrogencarbonate forms and this reacts with the sodium chloride to form sodium hydrogencarbonate which, being insoluble under the prevailing conditions, separates as a white solid which is continu-

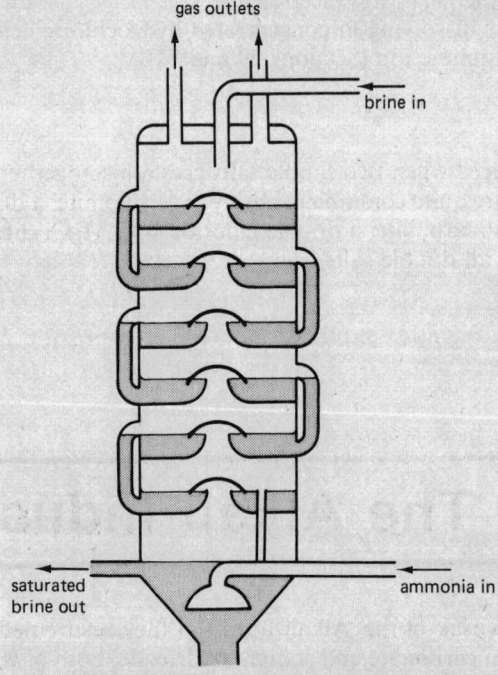

Figure 13   Ammonia scrubber

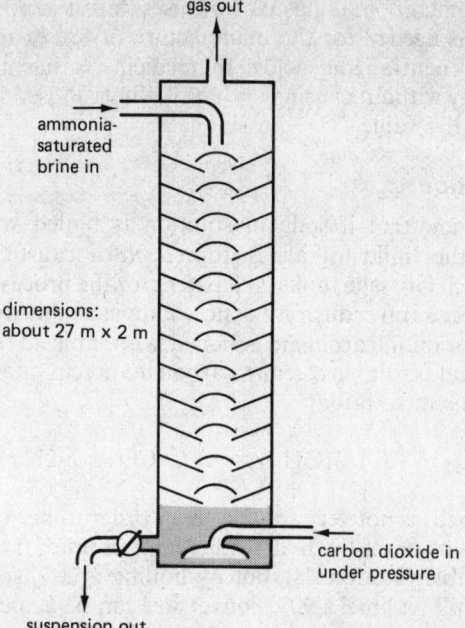

Figure 14   Solvay (or 'carbonating') tower

ously removed by rotary vacuum filters. The main reactions that take place in the Solvay towers are:

$$NH_3(g) + CO_2(g) + H_2O(l) \longrightarrow NH_4^+(aq) + HCO_3^-(aq)$$
$$NH_4^+ HCO_3^- + Na^+ Cl^- \longrightarrow NaHCO_3(s) + NH_4^+(aq) Cl^-(aq)$$

Finally, the hydrogencarbonate is heated to convert it to the carbonate:

$$2NaHCO_3 \longrightarrow Na_2CO_3 + H_2O + CO_2 \text{ (to carbonating towers)}$$

A feature of the ammonia-soda process is the maximum utilization of all the reagents and reactants, with the exception of the final calcium chloride. Heating the limestone produces calcium oxide and carbon dioxide that goes to the carbonating towers. The calcium oxide is slaked and then heated with the ammonium chloride, produced as above, to provide ammonia for the ammonia scrubber:

Limestone $\quad CaCO_3 \longrightarrow CaO + CO_2$ (goes to the carbonating towers)
Oxide slaked $CaO + H_2O \longrightarrow Ca(OH)_2$ (heated with ammonium chloride)
$\quad 2NH_4Cl + Ca(OH)_2 \longrightarrow CaCl_2 + 2H_2O + 2NH_3$
$\qquad\qquad\qquad\qquad$ (to ammonia scrubber)

Calcium chloride has some uses, e.g. refrigeration, soil consolidation, in the building and chemical trades, but most of it goes into the sea.

Sodium carbonate is usually transported in anhydrous form, because so much of the mass of washing soda is water (63%). Also, because the decahydrate is efflorescent (Section 30) the anhydrous salt can be much purer.

Sodium hydrogencarbonate has its own uses, but it is not so important as the carbonate. It is needed for the manufacture of fire extinguishers, baking powders, and for neutralizing acids. In medicine it has the advantage of neutralizing acidity without causing undue alkalinity if used in excess, because its pH is close to the value 7.

Sodium hydroxide

The alchemists knew that if soda in solution is boiled with 'caustic lime' (calcium oxide) the 'mild' alkali is turned into 'caustic' alkali (sodium hydroxide). In 1853, Gossage took out a patent for the process on an industrial scale, but to the chemical industry it is now known as the 'lime-soda' process.

If a solution of sodium carbonate is boiled with 'milk of lime' ion aggregation takes place and insoluble calcium carbonate precipitates, leaving sodium and hydroxide ions in solution:

$$(Na^+)_2CO_3{}^{2-} + Ca^{2+}(OH^-)_2 \longrightarrow CaCO_3(s) + 2Na^+ + 2OH^-$$

Calcium hydroxide is not very soluble so, in order to keep its concentration high, a thick suspension of the hydroxide ('milk of lime') has to be used. The reaction is reversible (Section 48), but by boiling a 20% solution of sodium carbonate with 'milk of lime' a 90% conversion can be achieved. A large proportion of sodium hydroxide is now produced in this way.

ELECTROLYTIC PRODUCTION OF SODIUM HYDROXIDE: The electrolytic cell must be designed so that, when brine is electrolysed, the chlorine is kept away from the sodium hydroxide with which it would react.

(1) Mercury cathode cell: The cell (Figure 15) is in two parts. The first part contains a line of graphite anodes dipping into strong brine which passes slowly through the cell. A layer of mercury, about 1 cm deep, passes along the bottom of the cell in the same direction as the brine flow. The mercury layer acts as the cathode and sodium ions are discharged there, immediately forming an amalgam with the mercury. Chloride ions are discharged as atoms at the graphite anodes and form molecules of chlorine gas:

$$2NaCl \longrightarrow 2Na^+ + 2Cl^-$$

At the anode(+) $2Cl^- - 2e^- \longrightarrow 2Cl \longrightarrow Cl_2(g)$

At the cathode(−) $2Na^+ + 2e^- \longrightarrow 2Na \longrightarrow Na/Hg$ amalgam

The second chamber contains water and auxiliary cathodes of graphite or iron. Here sodium from the amalgam displaces hydrogen, hydroxide and sodium ions going into solution:

$$2Na + 2H.OH(l) \longrightarrow 2Na^+(aq) + 2OH^-(aq) + H_2(g)$$

When the hydroxide concentration reaches about 50%, the solution is evaporated and the residue melted in iron vessels. If the second compartment is below the first, the mercury is returned to the latter by means of an Archimedean screw. The chlorine is evolved from the brine compartment and the hydrogen from the water compartment so there is no danger of the gases coming together. In fact, the two gases are valuable by-products. Electricity is costly in this country and the initial cost of the mercury is very high.

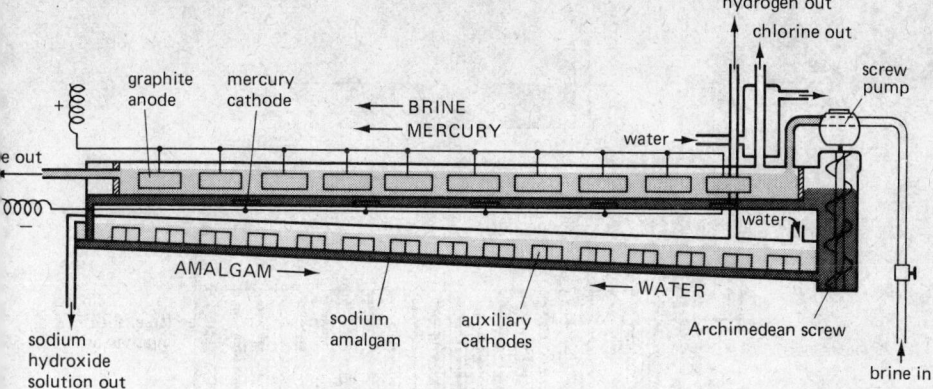

Figure 15   Mercury cathode cell

If chlorine is allowed to react with hydroxide ions it oxidizes them to hypochlorite ions:

$$2OH^- + Cl_2 \longrightarrow ClO^- + Cl^- + H_2O$$

Sodium hypochlorite, NaClO, is a valuable oxidizing and bleaching agent and this compound can be made by the action of chlorine on *cold* sodium hydroxide solution. The hypochlorite ion is not stable in hot solution because, on heating, chlorate and chloride ions are formed:

$$3ClO^- \longrightarrow ClO_3^- + 2Cl^-$$

Sodium chlorate can therefore be manufactured by passing chlorine into *hot* sodium hydroxide solution. Chlorates are powerful oxidizing agents and are poisonous. The sodium salt, $NaClO_3$, is deliquescent, but is used as a weed killer. All chlorates are dangerous compounds.

Summing up, this one important electrolytic process can produce not only sodium hydroxide but also hydrogen, chlorine, hydrogen chloride, hydrochloric acid, sodium hypochlorite, and sodium chlorate. Demand for chlorine varies from year to year and the decision to manufacture sodium hydroxide by ion aggregation or by electrolysis depends largely on this factor.

(2) Diaphragm cell: Many have tried to devise a cell for the electrolysis of brine that would be just as efficient but less costly than the mercury cathode cell. In most designs, the hydroxide ions are protected from the chlorine by a diaphragm of some kind that surrounds the cathodes. Diaphragms are often made of asbestos.

One of the most promising designs, the Hooker cell (Figure 16), was perfected in the U.S.A. between 1924 and 1934, by the Hooker Electrochemical Company at Niagara (where electric power is cheap). The cathodes, with their closely associated asbestos diaphragms, separate the caustic-containing electrolyte, that almost fills the cathode compartment, from the main body of the electrolyte. Large cells work with a current of 40 000 A and they are almost cubical in shape. Each cell has three main sections. The bottom part carries the anode assembly. On the concrete base is a lead casting that carries copper anode-connecting bars to the graphite anodes which alternate with, and are very close

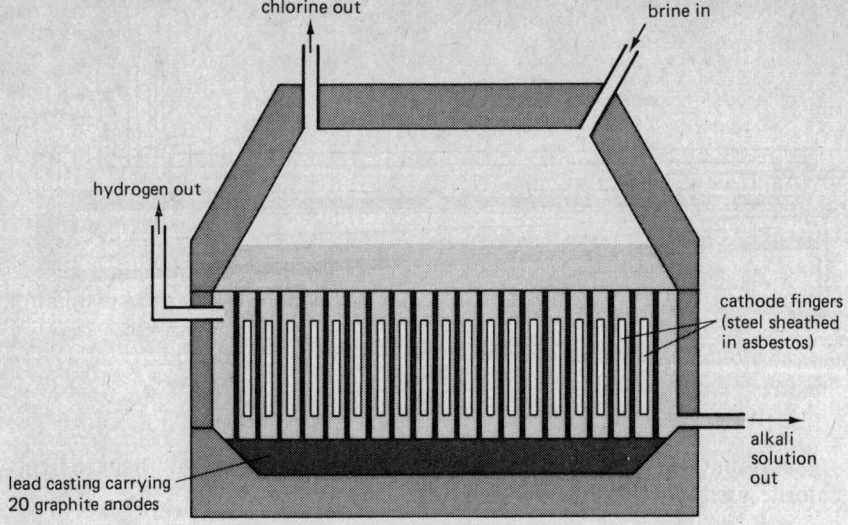

Figure 16   A Hooker cell

to, the vertical steel finger cathodes which are sheathed in asbestos. The middle part of the cell carries the cathode assembly which has a steel frame with copper conductor-bars to the steel cathodes. The top part, with its concrete lid, allows the entrance of fresh brine and the escape of chlorine. The tops of the anodes are well covered with brine.

Sodium hydroxide is required in large quantities for many purposes including purification of oils, absorption of acid gases, soap making, paper making, preparation of dyestuffs, manufacture of artificial silk, purification of aluminium oxide before the extraction of aluminium, etc.

39   Extraction and Chemistry of Some Common Metals: Sodium (Group I)

The best method for extracting any metal is dictated by economic considerations. Metals must be produced competitively and matters such as the cost of raw materials, relative cost of various fuels, cost of electricity, saleability of the by-products, removal of offensive by-products, necessity of refining, cost of transport, etc., have all to be considered. Therefore, what looks to be the best method from a study of chemical behaviour may not prove to be the best one economically. However, reference to the activity series given in Section 6 will give some guide.

From caesium to aluminium, sulphide or oxide ores are not common and most of the compounds of these elements are difficult to reduce by thermal methods. For these metals, therefore, electrolytic methods might be the answer and, in most cases, actually are. The four elements following aluminium are obtained by thermal processes – the reduction of oxide or sulphide ores – though much zinc is obtained by electrolytic methods. The next element, hydrogen, cannot be considered as a metal in this context because, being a gas, many extra problems are involved. Copper is obtained from sulphide ores (Section 72). Mercury can be obtained from its sulphide ore by a distillation process.

An oxide ore, for example of iron (Section 71) or tin, is usually reduced by heating with carbon. The overall reaction is exothermic and a high temperature is maintained. Carbon monoxide is produced and this gas is usually responsible for the greater part of the reduction. A sulphide ore cannot be reduced by heating with carbon because the formation of carbon disulphide, CS_2, is endothermic and no reducing compound, CS, (corresponding to CO) would form. The production of a metal, such as lead or zinc, from a sulphide ore, therefore, involves first the conversion of the sulphide to the oxide, by roasting in air, and then the reduction of the oxide with carbon.

In a few cases, an electropositive element can be used as the reducing agent. Examples: sodium in the preparation of titanium (see below); aluminium in the alumino-thermic reduction processes (Section 42). Hydrogen as a reducing agent has very limited use, though one important example is the reduction of tungsten oxide to prepare tungsten for the manufacture of electric-lamp filaments.

SODIUM (Group I)

In 1807, Davy first isolated sodium by electrolysis of the fused hydroxide. As the hydroxide now has to be manufactured from the chloride, which is cheap and plentiful (Section 38) it is more economical to obtain the metal directly by electrolysis of the fused chloride. This melts at a much higher temperature than the hydroxide (800 °C as against 318 °C) and at this temperature chlorine would react violently with the sodium hydroxide. The Downs cell, produced in 1924, overcomes most of the problems (Figure 17). The melt consists of 40% sodium chloride with 60% of calcium chloride, the presence of which lowers the melting point by about 200 °C. At this temperature (around 600 °C) sodium is preferentially discharged. The cell is made of steel with a refractory lining. The graphite anode is well covered with a suction hood for the safe collection of the evolved chlorine gas. From this hood hangs a cylindrical steel gauze and round this is the annular steel cathode. The sodium chloride has ions Na^+ and Cl^-.

At the cathode (reduction process) $Na^+ + e^- \longrightarrow Na$
At the anode (oxidation process) $Cl^- - e^- \longrightarrow Cl$

The chlorine atoms immediately pair to give Cl_2 molecules. Molten sodium rises from the cathode and collects outside the gauze.

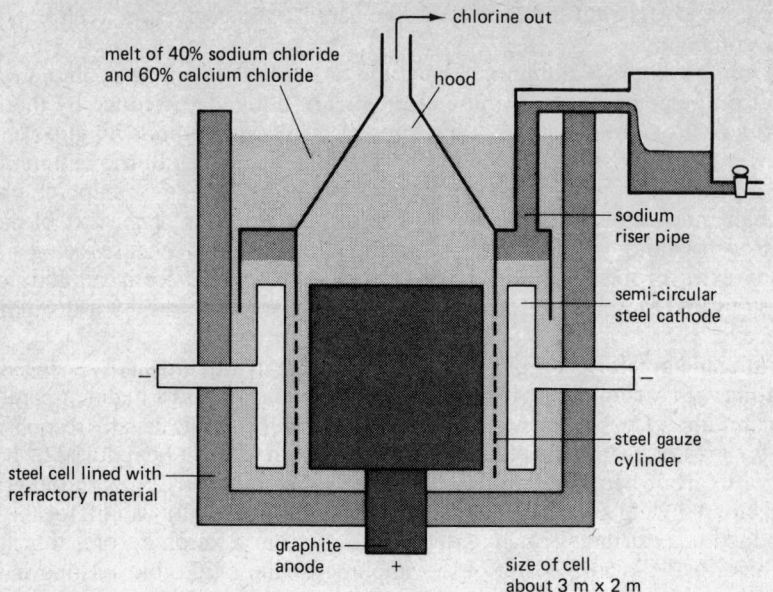

Figure 17　The Downs cell

Uses of sodium

Production of titanium metal; manufacture of anti-knock additives for petrol; manufacture of important compounds such as sodium peroxide, Na_2O_2; coolant in nuclear reactors; sodium lighting.

Titanium is obtained by the reduction of titanium(IV) chloride with sodium in an atmosphere of argon:

$$TiCl_4 + 4Na \longrightarrow 4NaCl + Ti$$

(Magnesium can also be used in place of sodium.) Titanium metal is nowadays quite common. Its compounds are widespread in nature and the metal is much lighter than steel though very strong. Structural material is now made with titanium alloys or even titanium itself.

Physical and chemical properties of sodium (see Section 22.)

40 Extraction and Chemistry of Some Common Metals: Calcium (Group II)

This element was first isolated by Davy in 1808. It was formerly produced by the electrolysis of the fused chloride but is now made by alumino-thermic reduction of pure calcium oxide in vacuo. The metal is not needed on a large scale, but it has some uses in dehydrating and as a constituent of some alloys. It is a very light metal (relative density 1.55 at 20 °C) which burns readily in air (bright-red flame) to form the oxide CaO (quicklime). It reacts slowly with cold water, but vigorously with hot, to give the hydroxide $Ca(OH)_2$ (slaked-lime) and hydrogen. The hydroxide is somewhat soluble in water giving an alkaline solution (lime-water). In all its compounds calcium is divalent.

Calcium carbonate, $CaCO_3$

Occurs widely in nature in many forms, such as (in descending order of purity) calcite (calc-spar or the transparent form – Iceland-spar), marble, chalk, lime-stone. It frequently occurs with magnesium carbonate, as in dolomite, $CaCO_3, MgCO_3$. Calcium and magnesium ions cause hardness in water (Section 27). Pyrolysis of calcium carbonate is a dissociation, not a decomposition as the reaction is reversible. If, however, it is carried out at 1000 °C, and the carbon dioxide is removed, the carbonate can be completely decomposed into the oxide (quicklime) and carbon dioxide:

$$CaCO_3(s) \rightleftharpoons CaO(s) + CO_2(g)$$

The decomposition is carried out in special lime-kilns. The resulting oxide is known as quicklime because of its lively reaction with water. If cold water is dripped onto freshly made oxide, there is a vigorous exothermic reaction with much swelling and steaming. If more water is added and the mixture filtered, a dilute solution of the hydroxide (lime-water) is obtained:

$$CaO + H_2O \longrightarrow Ca(OH)_2$$

When lime-water comes into contact with carbon dioxide, the carbonate is again formed. The lime-water turns turbid (test for carbon dioxide):

$$Ca(OH)_2(aq) + CO_2(g) \longrightarrow CaCO_3(s) + H_2O$$

If more carbon dioxide is passed through a turbid lime-water suspension the liquid eventually clears, due to the divalent carbonate ions being converted to monovalent hydrogencarbonate (bicarbonate) ions:

$$CO_3{}^{2-} + H_2O + CO_2 \rightleftharpoons 2HCO_3{}^-$$

This reaction is reversible and, if the clear solution is heated, a white precipitate of calcium carbonate is seen again.

Builder's mortar is made by slaking quicklime and mixing it with about three times as much sand. The slaked lime hardens and the sand prevents the mortar from cracking. Nowadays, cement is in much commoner use than mortar and this is made by heating marl in huge rotating cylinders. Marl is a mixture of limestone and clay; natural marl contains about two-thirds limestone to one-third clay. Cement is a complicated mixture of silicates and aluminates of calcium and, when water is added, it crystallizes slowly into an extremely strong structure of interlacing needles. It is often known as Portland cement because of its resemblance to natural Portland stone. Cement mixed with gravel or ballast produces concrete.

Calcium sulphate, $CaSO_4$

This occurs in nature either as the dihydrate 'gypsum', $CaSO_4.2H_2O$, or as 'anhydrite', $CaSO_4$. It is only slightly soluble in water but it can cause permanent hardness (Section 27). When gypsum is heated to 120 °C it loses three-quarters of its water forming the 'hemihydrate', $CaSO_4.\frac{1}{2}H_2O$ (alternatively formulated as $(CaSO_4)_2,H_2O$). This is 'Plaster of Paris' which, when water is added, forms gypsum again with the evolution of heat and the compound sets hard with a slight increase in volume. It is commonly used for making plaster casts, for encasing limbs after the setting of fractures and for wall surfaces. Blackboard 'chalk' is often a preparation from gypsum. It does not react with dilute acid.

Calcium chloride, $CaCl_2$

Calcium chloride can be obtained by adding excess calcium carbonate to dilute hydrochloric acid, filtering, and evaporating the filtrate to crystallizing point. Crystals of the hexahydrate, $CaCl_2,6H_2O$, are obtained. There is no hydrolysis (compare with magnesium, Section 41). The chloride is obtained as an end-product of the ammonia-soda process (Section 38). When heated strongly, hydrated calcium chloride loses its water and then melts. The anhydrous (fused) compound is a laboratory drying-agent, often used in desiccators.

Calcium fluoride, CaF_2

Calcium fluoride, CaF_2, occurs in nature as fluorite ('fluor-spar'). It is the source of hydrogen fluoride (and hence fluorine) and is also used industrially as a flux (for lowering melting points). A very little, when present in drinking water, arrests tooth decay.

41   Extraction and Chemistry of Some Common Metals: Magnesium (Group II)

Magnesium was probably first discovered by Davy but it was not immediately recognized as a new element. Bunsen, in 1852, was the first to obtain it by electrolysis using the fused chloride. The chloride can now be obtained from sea water, which contains small amounts of magnesium chloride and magnesium sulphate together with many other salts; sodium chloride in particular. To separate the magnesium, its salts are first precipitated as the hydroxide by adding calcium hydroxide. The precipitate is then separated and heated to form the oxide, which is then mixed with carbon and heated in a stream of chlorine:

$$MgO + C + Cl_2 \longrightarrow MgCl_2 + CO$$

The fused chloride is electrolysed with carbon anodes and steel cathodes (Figure 18). Sodium chloride and calcium chloride are also present in the melt but magnesium is preferentially discharged at the cathodes. The specially designed cell has 'semi-walls' (or 'curtain plates') made of refractory material between the anodes and cathodes to keep the chlorine away from the liberated metal. Most of the metal used in this country is imported from Germany and elsewhere.

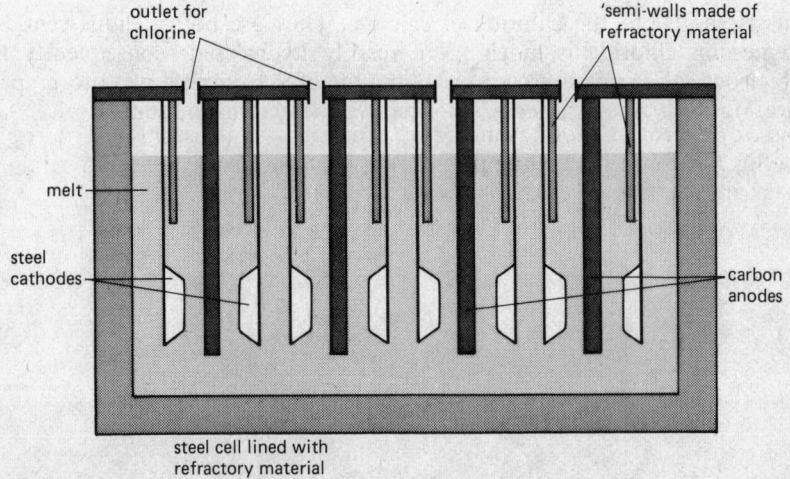

Figure 18   Electrolytic extraction of magnesium

Magnesium is denser than calcium but less dense than aluminium, its relative density being 1.74 at 20 °C. From its position in the Periodic Classification one would expect magnesium to have properties similar to those of calcium but to be rather less electropositive. The element, like calcium, is always divalent in its compounds and these are ionic in character. Magnesium compounds are widespread in nature and often associated with those of calcium. Magnesium reacts with water much less readily than calcium does but it burns brightly in steam and in many other gases, including nitrogen and carbon dioxide, mainly because the temperature is so high. It melts at about 650 °C and a piece of magnesium ribbon held in a hot flame is seen to melt before it ignites. A block of the metal does not ignite easily.

The metal has a number of uses, particularly in powder form, including pyrotechny ('fireworks'), flash-light powders, reducing agent for the production of titanium or silicon, manufacture of some light alloys.

Magnesium carbonate, $MgCO_3$

This has properties similar to those of calcium carbonate. Although insoluble in water, it will form hydrogencarbonate ions when carbon dioxide is present, just as calcium carbonate does. When heated strongly, the oxide, MgO, is formed and as this has a very high melting point (2800 °C) it is a very useful refractory material. Unlike calcium oxide, it is almost insoluble in water and a film of oxide on the surface of the metal can act as a protective coating. A strip of magnesium ribbon may, in consequence, look dull in appearance but the film is easily removed with glass paper.

Magnesium sulphate, $MgSO_4$

This is quite soluble in water (unlike calcium sulphate) and it also causes permanent hardness in water. Its hydrated form, $MgSO_4,7H_2O$, occurs in nature as 'Epsomite' (Epsom salt). It has important uses in industry for weighting and fire-proofing of fabrics, for dyeing and in medicine as a mild purgative.

Magnesium chloride, $MgCl_2$, $6H_2O$

This is similar to the chloride of calcium. They are both deliquescent but magnesium chloride is much more readily hydrolysed; consequently the anhydrous salt is not so easy to obtain. One of the simplest ways to prepare pure $MgCl_2$ is by passing dry chlorine over heated magnesium.

42 Extraction and Chemistry of Some Common Metals: Aluminium (Group III)

Aluminium was first isolated in 1827 in impure form, but it was not until 1886 that Hall, in the U.S.A., and Héroult, in France, working quite independently, found an efficient way of preparing the metal by electrolysis. Aluminium compounds are common in nature but the element usually occurs in complicated silicate minerals from which it is difficult to obtain the metal. The oxide and chloride of aluminium are covalent compounds that do not conduct electricity when fused. Aluminium oxide is a very difficult oxide to reduce. It is also difficult to refine crude aluminium and so the starting materials have to be pure.

The Hall-Héroult process

The starting material is the oxide mineral called bauxite, which now comes mainly from Jamaica. To purify this ore (the Bayer process) the ground ore is digested under pressure with a hot solution of sodium hydroxide (10%–30% concentration). The aluminium oxide, being amphoteric in nature, dissolves as sodium aluminate:

$$Al_2O_3 + 2NaOH \longrightarrow 2NaAlO_2 + H_2O$$

The solid residue is filtered off and when the filtrate is cool it is 'seeded' with a few crystals of aluminium oxide trihydrate, $Al_2O_3,3H_2O$. This causes most of the aluminium to crystallize as hydrated oxide which is separated and strongly ignited to obtain pure anhydrous oxide.

The purified oxide is dissolved in molten cryolite and electrolysed. Fluorite, CaF_2, may also be added to lower the melting point. Cryolite, Na_3AlF_6, used to be obtained in its natural state from Greenland, but now it is prepared synthetically. Molten cryolite can dissolve about 15% aluminium oxide at 1000 °C and the mixture conducts well. Normally the cell works with about 5% oxide. It is difficult to discover how the cryolite plays its part, but the final result is the decomposition of the oxide with the cryolite remaining unchanged, although there is no doubt that it is chemically involved as an intermediary. Aluminium appears at the cathode and oxygen at the anode:

At the cathode $(-)$ $4Al^{3+} + 12e^- \longrightarrow 4Al$
At the anode $(+)$ $6O^{2-} - 12e^- \longrightarrow 6O \longrightarrow 3O_2$

Overall decomposition $2Al_2O_3 \longrightarrow 4Al + 3O_2$

The carbon block anodes are immersed in the melt and kept about 5 to 8 cm above the molten aluminium which collects on the floor of the cell. The cell has a steel base and walls lined with carbon (about 30 cm thick) which acts as the cathode. When the concentration of the aluminium oxide has dropped from 5% to about 1%, more oxide is added. The current may be 40 000 A, the voltage 5 V, and, in a large plant, there may be as many as 2000 cells connected in series.

The carbon-block anodes are made by moulding good coke and hard pitch under pressure. They gradually burn away in the oxygen that forms on them (about 3 cm per day) giving carbon dioxide with a little monoxide. To save the expense of stopping the cells' operation in order to replace consumed anodes, the Soderberg type (Figure 19) are so made that the anode paste can be fed in

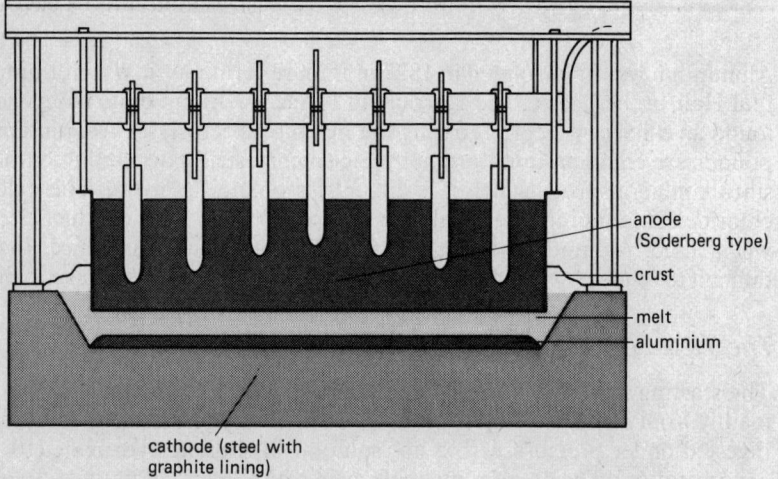

anode
(Soderberg type)

crust

melt
aluminium

cathode (steel with
graphite lining)

Figure 19 Electrolytic extraction of aluminium

as the cell operates to replace that which is burnt away, the paste being baked hard by the heat of the cell (operating temperature about 1000 °C). External heating is not required as the initial mixture is melted electrothermally.

Aluminium is trivalent in all its common compounds and the compounds themselves are mainly covalent. It is a comparatively light metal, though its relative density, 2.7 at 20 C, is higher than those of the three metals so far considered, sodium, calcium and magnesium. In air, aluminium gains a dull coating of oxide which makes its reactivity with acids, etc., rather less than would be expected from its position in the activity series. Metals of Groups I and II also form coatings of oxide but these are ionic compounds and, with the exception of magnesium oxide, are soluble in water. Aluminium oxide, being covalent and insoluble, protects the metal from the action of some reagents. If aluminium is made the anode of an electrolytic cell containing a strong oxidizing agent, such as chromic acid (potassium dichromate with dilute sulphuric acid), the oxide layer can be thickened by 'anodic oxidation'. This layer can be coloured with suitable dyes, to make very attractive finishes known as **'anodized' aluminium.**

The protective oxide layer affects the action of acids with aluminium. Nitric acid, at any concentration, has hardly any effect on the metal. As this acid is a powerful oxidizing agent, it simply thickens the oxide layer. Dilute sulphuric acid has little action but the hot concentrated acid breaks down the protective layer and reacts with the metal. Hydrochloric acid reacts at most concentrations and the solution formed contains hydrated aluminium ions, $Al(H_2O)_6^{3+}$, and chloride ions.

Aluminium has an amphoteric nature; its oxide and hydroxide are both amphoteric (Section 32). The metal reacts readily with strong alkalis, aluminate ions and hydrogen being formed:

$$2Al + 2OH^- + 2H_2O \longrightarrow 2AlO_2^- + 3H_2$$

Sodium carbonate solution has a similar reaction when warmed with aluminium powder (hydroxide ions being produced by the hydrolysis of the carbonate). Therefore aluminium cooking utensils should not be cleaned with soda. Such utensils are normally protected by the oxide film but some acids and alkalis can remove it. The protective film can be removed in a spectacular way by rubbing the surface of aluminium with a little moist mercury(II) chloride (poisonous!) when the aluminium oxidizes rapidly at the exposed surface with the evolution of heat.

Aluminium has many uses mainly because of its strength, lightness, good conductivity, and inertness to reagents (due to the oxide layer). A metre of aluminium wire with the same conductance as a metre of copper wire will be thicker but weigh less than the copper wire. Copper is about three times as dense as aluminium but its conductance is not three times as great. Rolling stock made of aluminium does not need the expense of regular painting that steel has to have. If the surface of the aluminium is grazed, the air soon restores the protective oxide film.

The chief uses of aluminium are: aircraft construction, railway rolling stock, electric cables, building materials (e.g. roofing), liquid containers (e.g. beer barrels), packaging, chemical plant, ships' superstructures, domestic utensils, machinery of all kinds where weight saving is essential, and in '**alumino-thermic' reduction** ('Thermit(e) Process').

Oxides that are difficult to reduce, such as iron(III) oxide, Fe_2O_3, chromium(III) oxide Cr_2O_3, and the compound oxide of manganese, Mn_3O_4, can be reduced to the metal by reaction with aluminium powder. The oxide and the powder are well mixed and the reaction, which will not start at low temperatures, is initiated by igniting a small amount of an aluminium/barium peroxide mixture. The latter reacts vigorously, produces a high temperature, and starts the main reaction. Aluminium oxide has a very high heat of formation (Section 46), about $-1600\,kJ\,mol^{-1}$, so this 'alumino-thermic' reduction is strongly exothermic. With iron(III) oxide the reaction is:

$$Fe_2O_3 + 2Al \longrightarrow Al_2O_3 + 2Fe \qquad \Delta H = -770\,kJ$$

(For the significance of ΔH see Section 46.)

Pure aluminium is seldom encountered in everyday life. For most purposes an alloy is used and there is now an enormous variety of these. Two of the earliest were 'duralumin' (Al, 95; Cu, 4; Mn, $\frac{1}{2}$; Mg, $\frac{1}{2}$) which, as the name suggests, is a harder metal and 'magnalium' (Al, 70; Mg, 30), the figures in parentheses being percentages by mass.

Aluminium oxide, Al_2O_3

Insoluble in water but amphoteric, aluminium oxide slowly reacts with any common strong acid or alkali to give the corresponding salt. Some gem stones, e.g. ruby, consist of it and it is now used in the preparation of artificial gems. Its melting point is very high (2050 °C) and crucibles that will function at high temperatures are made from it. The specially prepared powder is used in chromatography.

Aluminium hydroxide, $Al(OH)_3$

This is precipitated as a gelatinous white substance when a hydroxide solution (sodium, potassium or ammonium) is added to a solution of an aluminium salt, e.g. the sulphate:

$$Al^{3+} + 3OH^- \longrightarrow Al(OH)_3$$

As it is amphoteric, the precipitate redissolves in an excess of either of the first two alkalis mentioned but *not* in an excess of ammonium hydroxide. Aluminium hydroxide is used in industry as a **mordant** in dyeing (i.e. providing a medium that will adsorb colouring matter so that it will not wash out), in treatment of the water supply, and in the treatment of sewage. On ignition, the hydroxide gives the pure oxide:

$$2Al(OH)_3 \longrightarrow Al_2O_3 + 3H_2O$$

Aluminium sulphate, $Al_2(SO_4)_3$

This commonly encountered compound of aluminium is obtained industrially by the action of sulphuric acid on clay or bauxite. One mole of aluminium sulphate crystallizes with 18 moles of water. The sulphate is very soluble and, like the chloride, is easily hydrolysed.

It is used in the manufacture of 'foam' fire-extinguishers. These make use of the reaction between aluminium sulphate solution and sodium hydrogencarbonate, saponin also being added. The hydrolysis of the sulphate produces sulphuric acid which reacts with the hydrogencarbonate to give a multitude of bubbles – carbon dioxide imprisoned in the foam of aluminium hydroxide. Saponin (obtained from horse chestnuts) encourages the production of foam and gives it lasting quality.

Industrially, aluminium sulphate is usually used in the form of the double sulphate with potassium. This is known as 'potash alum' and is easily made by mixing hot concentrated solutions of each sulphate and allowing the mixture to cool. The alum is much more soluble in hot water than it is in cold and thus can be purified easily by simple recrystallization. The double salt does *not* hydrolyse. The name **alum** has now become a general term for a double sulphate formed from any monovalent metal (or ammonium) and the sulphate of *any* trivalent metal. The mole of any alum crystallizes with 24 moles of water. Common examples include:

Potash alum (colourless) $K_2SO_4, Al_2(SO_4)_3, 24H_2O$
Chrome alum (purple) $K_2SO_4, Cr_2(SO_4)_3, 24H_2O$
Iron alum (pale mauve) $(NH_4)_2SO_4, Fe_2(SO_4)_3, 24H_2O$

Aluminium chloride, Al_2Cl_6

The molecule $AlCl_3$ would be deficient in electrons and able to accept a pair. This explains why aluminium chloride exists as double molecules and why it can easily form an addition compound with another covalent compound that has a 'lone pair' of electrons to offer, such as ammonia (Section 10):

$$
\begin{array}{ccc}
:\ddot{C}l: & & H \\
:\ddot{C}l:\overset{..}{Al} & \longleftarrow & :\overset{..}{N}:H \\
:\ddot{C}l: & & H
\end{array}
$$

It is an important catalyst in some organic processes and its action is probably explained by its readiness to accept electrons. Aluminium chloride is so easily hydrolysed that it fumes in moist air and this is due to the slow evolution of hydrogen chloride which forms a mist of hydrochloric acid:

$$Al_2Cl_6(s) + 6H_2O(l) \longrightarrow 2Al(OH)_3(s) + 6HCl(g)$$

43 Extraction and Chemistry of Some Common Metals: Tin (Group IV)

Tin is a rare and valuable metal that is found as tin(IV) oxide, SnO_2, in small quantities, mainly in S.E. Asia. Cornwall, England, still has one or two mines in operation which may, because of the high price of tin, be further exploited. Tin is extracted by a method similar to that used for iron (Section 71). The concentrated ore is mixed with coke and a little crushed limestone and heated in a blast furnace. Carbon monoxide is formed and this is the chief reducing agent:

$$SnO_2 + 2CO \longrightarrow Sn + 2CO_2$$

The symbol Sn for the atom of tin is derived from its Latin name 'stannum' and it was the Romans who first developed the Cornish tin mines. The metal has a relatively low melting point of 232 °C and its relative density (7.3) is about the same as that of zinc (7.1). The element has three allotropic forms (Section 50).

Variable valence is expected as a typical property of the transition elements (Section 70), but in Group IV both tin and lead have valencies of two and four.

One would expect the valence to be four only, because each element has atoms with four valence electrons. One electron pair, however, is often inactive and is known as an 'inert pair'. In the case of both tin and lead such electron inertness can result in the valence of two. The divalent compounds of both these elements are ionic whereas the tetravalent ones are covalent.

Tin, because it is only just above hydrogen in the activity series is little affected by air, water, dilute acids or alkalis. It retains its brightness in air and is much used for the making of tin plate, i.e. iron that has been coated with tin in order to prevent rusting. Because tin is so expensive tin plate is usually made by electrolytic methods rather than by the older 'hot-dip' process. Electrolytically a thinner and more even coating can be applied. As tin is below iron in the activity series, should the coating become broken the iron dissolves and not the tin. This is a great advantage for canning foodstuffs because iron in food would not be nearly as dangerous as small quantities of tin. (Compare this with 'galvanized iron', in Section 45.) Other uses of tin include the preparation of low-melting alloys (such as pewter, solder, type metal, etc. (Section 44)) and as foil. Tin foil, however, is costly and has largely been replaced by aluminium for food wrapping, etc.

Chemically, an interesting reaction of tin, and one by which the metal can be recognized, is its behaviour with concentrated nitric acid. All common nitrates are soluble in water and when a metal reacts with nitric acid to form a nitrate no precipitate is seen. Tin, however, is oxidized directly to tin(IV) oxide and a white precipitate appears when the mixture is warmed and this precipitate does not dissolve when diluted with water. This is a remarkable example of the oxidizing power of nitric acid.

The most important tin(IV) compound is the oxide SnO_2. It is used for making white-glazed tiles, milk-white glass, as a polishing powder, etc. The most important tin(II) compound is the chloride $SnCl_2$ which, because of its readiness to lose two electrons to become tin(IV), acts as a powerful reducing agent. It will reduce solutions of iron(III) compounds to the iron(II) state and copper(II) compounds to copper(I).

No carbonate of tin is known.

44   Extraction and Chemistry of Some Common Metals: Lead (Group IV)

There is far more lead in the Earth's crust than tin and most of it is obtained from sulphide ores such as galena (lead(II) sulphide), PbS. This ore is usually found with zinc blende (zinc(II) sulphide), ZnS, so the metallurgies of these two metals are closely intertwined. Having separated the galena from the zinc blende, the sulphide ore has to be oxidized on a sintering machine (see Figure 20). (To sinter means to raise a crushed solid to such a temperature that the particles begin to melt and cohere.) When cooled, the product is a porous conglomerate rather like the 'clinker' that forms in a domestic solid-fuel boiler. In this case, the sintering machine not only produces the required sinter but oxidizes the sulphide to oxide at the same time:

$$2PbS + 3O_2(air) \longrightarrow 2PbO + 2SO_2$$

The sulphur dioxide is used for the production of sulphuric acid.

The sintered lead(II) oxide ('litharge') is mixed with limestone and coke and transferred to a blast furnace. Here the oxide is reduced by the coke (and carbon monoxide formed from it) to lead. This part of the operation is similar to that used in the extraction of iron (Section 71). Molten lead is withdrawn from the base of the furnace. Lead extracted in this way contains a small amount of silver which is worth recovering and the 'desilverization' of lead is of major importance. In all processes dealing with lead, strict precautions have to be observed because of its poisonous nature.

Special blast furnaces are now in operation that extract both lead and zinc from sulphide ore that contains both. These obviate the necessity of separating the two sulphides and are more economical (Section 45).

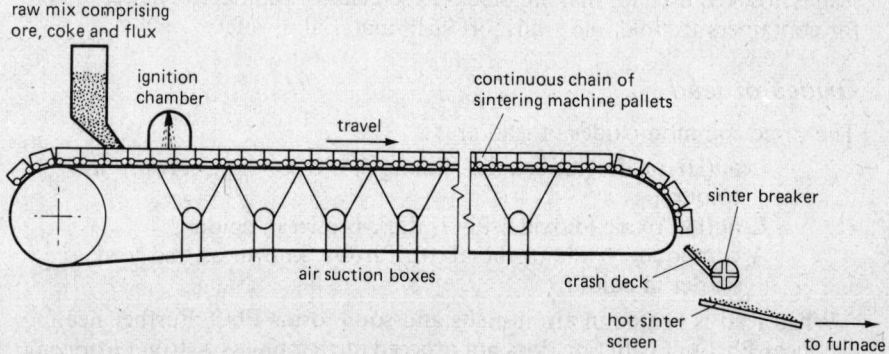

Figure 20   A sintering machine

As with tin, the symbol Pb for an atom of lead is derived from its Latin name 'plumbum' and it was the Romans who exploited lead ores in Britain – though the mines are now derelict. Lead is a relatively heavy metal (relative density 11.3 at 20 °C). It is a soft metal and marks paper when rubbed on it. This has caused confusion with graphite (Section 49). Like barium and mercury, lead is classed as a 'heavy metal' because of its high atomic number, 82. Compounds of 'heavy' metals are likely to be (a) insoluble in water and (b) poisonous in the human system. The human system has no way of eliminating them and small quantities of lead can add up as a 'cumulative poison' until serious trouble is caused.

Like tin, the atoms of lead have an inert pair of electrons, so there can be lead(II) compounds that are ionic and lead(IV) compounds that are covalent. In the laboratory, lead(IV) compounds, with the exception of PbO_2, are seldom encountered but there are industrially important compounds, such as tetraethyl lead (TEL) which is used as an 'anti-knock' additive to petrol. Of the common lead(II) compounds, only the ethanoate (acetate) and nitrate are readily soluble in water. All are very poisonous. As with tin, lead is little affected by air, water, dilute acids or alkalis. With cold concentrated sulphuric acid, a protective sulphate film is formed, but hot acid above 70% concentration can break down this film and then the metal reacts readily with the evolution of sulphur dioxide. Hot concentrated hydrochloric acid reacts only slowly with lead, but nitric acid at most concentrations forms the nitrate with the evolution of oxides of nitrogen (these two reactions being unlike those of tin).

Lead is appreciably soluble when warmed with distilled water and if 'soft' water is passed through lead pipes for drinking there may be danger. Water that is sufficiently 'hard' (Section 27) soon forms a protective coating of basic carbonate and/or sulphate. On exposure to air, bright lead foil soon becomes tarnished with a film of basic lead(II) carbonate. Its melting point of 327 °C is not as low as that of tin, but both metals are used for the preparation of low-melting alloys.

Formerly the chief uses of lead were: roof covering, sheathing electric cables, paint manufacture, alloys. For the first of these copper, zinc, and aluminium are becoming more popular as lead becomes more costly. For the second, the plastic PVC (Section 57) has replaced it except for the covering of large main cables and for the third much less lead is used because of its poisonous nature. However, other important uses for the metal have been found, chiefly in the motor-car industry for accumulators and petrol additives. A large quantity of lead is now required for making blocks as screens for radioactive materials and for containers to hold and transport such material in safety.

Oxides of lead

The three common oxides of lead are:

>Lead(II) oxide (monoxide, 'litharge') PbO_2, light brown in colour;
>Lead(IV) oxide (dioxide) PbO_2 dark-brown in colour;
>A compound oxide of the above Pb_3O_4, known as 'red lead', scarlet in colour.

When lead is heated in air, it melts and soon forms PbO. Further heating produces Pb_3O_4. Oxidation does not proceed further unless a strong oxidizing agent is used. If the red oxide is heated with a little dilute nitric acid, almost immediately a colour change from red to dark brown is seen. Only one-third of the lead in 'red lead' is converted to the dioxide, the other two-thirds producing the nitrate:

$$Pb_3O_4 + 4HNO_3 \longrightarrow 2Pb(NO_3)_2 + PbO_2 + 2H_2O$$

Lead(II) oxide is easily obtained in the laboratory by pyrolysis of lead(II) carbonate or nitrate or one of the other oxides. It is mainly basic in character and can be reduced to the metal by heating in a reducing gas (hydrogen, ammonia or carbon monoxide) or with carbon on a charcoal block. It is used in glass manufacture and in pottery glazing. Lead(IV) oxide is very important in the manufacture of the positive plates of lead accumulators. It is a powerful oxidizing agent and readily liberates chlorine when warmed with concentrated hydrochloric acid:

$$PbO_2 + 4HCl \longrightarrow PbCl_2 + 2H_2O + Cl_2$$

It is an amphoteric oxide but is *not* a peroxide (Section 32).

The compound oxide (red lead), Pb_3O_4, is a pure compound though it is convenient to regard it as $(PbO)_2.PbO_2$. It has the properties of both these oxides. It gives oxygen on heating and becomes entirely PbO. It also acts as an oxidizing agent liberating chlorine from hydrochloric acid. Commercial red lead is used in glass manufacture and always contains some lead(II) oxide. Its use with 'white lead' (see below) as a paint primer is much less important now that modern non-toxic preparations are available.

Lead(II) 'hydroxide'

This is obtained as a gelatinous white precipitate when an alkaline hydroxide solution is added to the solution of a lead(II) salt, e.g. the nitrate. The precipitate is probably a hydrated form of lead(II) oxide with the possible formula $(PbO)_2,H_2O$.

Lead(II) carbonate

This is usually encountered in the form of 'white lead' which is a basic carbonate – partly $PbCO_3$ and partly hydrated oxide. White lead has great covering power and was formerly used on a large scale as the basis of white paints and, in conjunction with various pigments, coloured paints as well. It has now been largely superseded by titanium(IV) oxide, TiO_2, which is non-toxic and is not blackened by hydrogen sulphide.

Lead(II) ethanoate (acetate), $Pb(CH_3.COO)_2$

This is one of the few lead salts that are readily soluble in water. It is used in the laboratory as a test for the presence of hydrogen sulphide because, being the salt of a weak acid, it reacts more readily with hydrogen sulphide (itself a weak acid) than the nitrate does, to produce black lead(II) sulphide:

$$Pb(CH_3.COO)_2(aq) + H_2S(g) \longrightarrow PbS(s) + 2CH_3.COOH(aq)$$

It can be obtained by dissolving lead(II) oxide or carbonate in warm dilute ethanoic (acetic) acid.

Lead(II) chloride, PbCl₂

This is almost insoluble in cold water but dissolves fairly easily in hot. Therefore it can readily be purified by recrystallization. If a small amount of the solid is heated with distilled water it will disappear but, on cooling, pure crystals of the chloride precipitate. Lead(II) chloride is less soluble in cold dilute hydrochloric acid but in concentrated hydrochloric acid it dissolves readily because of the formation of complex ions $[PbCl_4]^{2-}$ (Section 37).

Lead(II) sulphate, PbSO₄

This is also much more soluble in hot water than it is in cold.

Alloys of tin and lead

An alloy results when a metal has its properties modified by adding to it one or more other elements. The added substances are usually also metals, but this is not always so. Carbon, for example, is important in steel (Section 71) and phosphorus in phosphor-bronze. Many alloys, like those made from tin and lead, can be prepared very simply by melting one metal and stirring into it the requisite amount of the other. What results is either a solid solution (Section 28), a chemical compound, or a mixture of both. The phase diagram in Figure 21 shows what happens in a simple case: lead stirred into molten tin or *vice versa*.

As the lead is stirred into molten tin the melting point of the tin is steadily lowered and, similarly, when tin is stirred into molten lead its melting point is

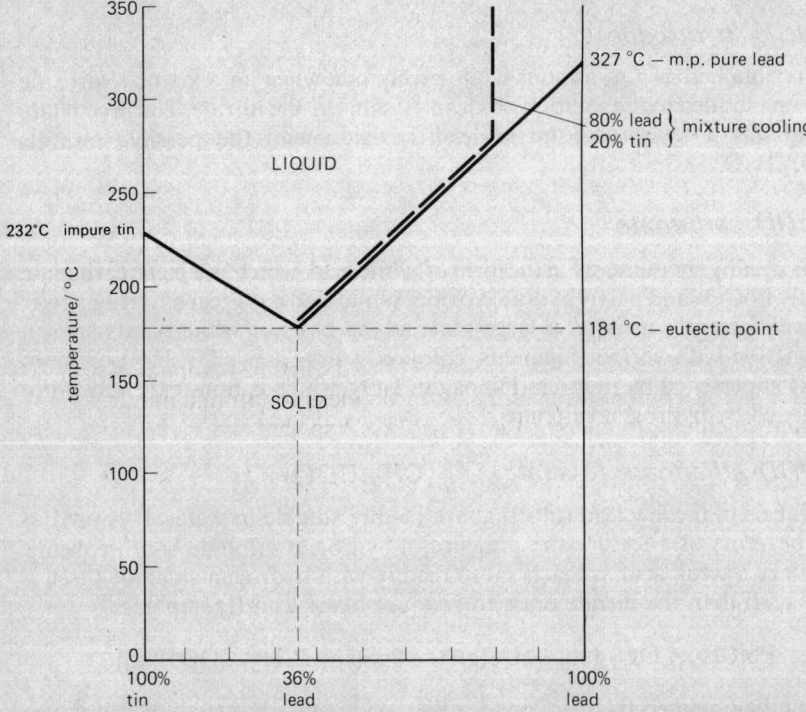

Figure 21　Phase diagram for mixtures of tin and lead

also lowered. In Figure 21, the two lines meet at what is known as the **eutectic point,** which represents the lowest temperature (181 °C) at which solid and liquid can be in equilibrium with each other. This is, as the diagram shows, when the composition of the alloy is 64% tin; 36% lead (by mass). If a molten mixture of 80% lead and 20% tin was made at 400 °C and allowed to cool slowly, then when the temperature reached the point on the line shown (somewhat below 300 °C) solid lead would separate continuously as the temperature dropped further. When 181 °C was reached the whole mass would solidify and the metal would contain the eutectic mixture with extra lead incorporated. No compound of tin and lead is formed.

Old pewter, prized by antique-collectors, contained 80% tin: 20% lead and, because of the high cost of tin, would now be expensive to make. 'Fine' solder contains about 66% tin (approximately the eutectic mixture), 'soft' solder about 50% of each metal, and cheap 'plumber's solder' only about 30% tin – though this has a useful 'pasty' range between 181 °C and 225 °C. 'Spotted metal' (used for some organ pipes) (35–65% tin) shows spots of lead on the surface. 'Type metal' is mainly lead with 3–11% tin and 11–20% antimony.

45 Extraction and Chemistry of Some Common Metals: Zinc (Transition)

In the Periodic Classification, zinc comes at the end of the first transition series. The transition elements (Section 70) have an unfilled penultimate (next to last) shell of electrons and their special properties derive from this. Zinc does not have these properties because, being the last of the series, the penultimate shell of electrons is completely filled. Zinc has atomic number 30 and its electron configuration is 2,8,18,2. Magnesium (atomic number 12) is 2,8,2 and calcium (atomic number 20) is 2,8,8,2 so that we might expect the properties of zinc to show similarities with those of magnesium and calcium. This is indeed the case and all three metals are divalent.

The metals that have been briefly discussed in Sections 39–44 can be isolated by either electrolytic methods *or* by thermal methods. Zinc is unusual in that about one-third of the world's production is by thermal means and about two-thirds by electrolytic means.

Thermal reduction

(1) VERTICAL RETORT PROCESS: The sulphide ore, zinc blende, is roasted and sintered on a sintering machine (Section 44) in much the same way as lead sulphide is the by-product, sulphur dioxide, being used for sulphuric acid manufacture:

$$2ZnS + 3O_2(air) \longrightarrow 2ZnO + 2SO_2$$

The sintered oxide is then made into briquettes with coal and heated to a temperature of about 1000 °C in vertical retorts. Carbon and carbon monoxide reduce the oxide to the metal:

$$ZnO + C \longrightarrow Zn + CO$$
$$ZnO + CO \longrightarrow Zn + CO_2$$
$$CO_2 + C \longrightarrow 2CO$$

Zinc, about 98% pure, distils from the mixture and, if purer zinc is required, it can be distilled again using a fractionating column.

(2) ZINC/LEAD BLAST-FURNACE: In 1950, at Avonmouth, a most unusual development took place. A blast furnace came into operation that could deal with a zinc/lead ore in one operation, producing zinc from the top and lead from the bottom. Not only does this obviate the necessity of separating the two ores but it also substitutes one fairly simple metallurgical process for two separate ones, thus greatly reducing cost. Since 1950 many more such furnaces have been put into operation in England, France, Australia, Zambia, etc.

Lead melts at 327 °C; zinc at 419 °C. Lead boils at 1750 °C; zinc at 907 °C. The hottest part of the furnace being about 1300 °C, only the zinc distils.

In the shaft of the furnace, metallic zinc and lead are produced by reduction with carbon and carbon monoxide (equations already given) (Figure 22). Molten lead goes to the bottom of the furnace and zinc vapour goes to the top. Here it is captured by a 'lead-splash condenser'. Sprays of molten lead

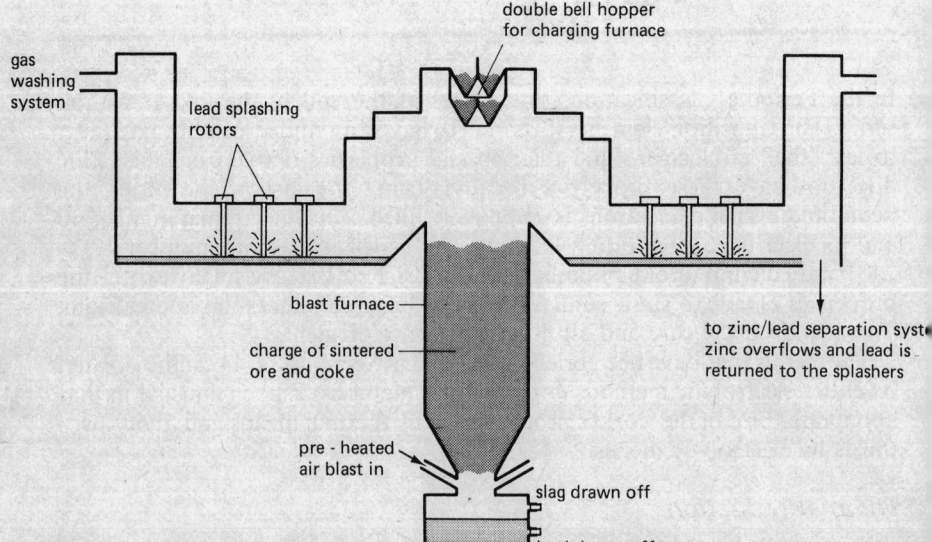

Figure 22   Blast furnace for the extraction of lead and zinc

droplets collect the zinc from the flue gases as an alloy and when this is removed and cooled it separates into two layers – the zinc at the top and the lead below. The zinc is removed and the lead returned to the lead splashers to collect more zinc. About 90% of the zinc is recovered and it is at least 98.5% pure. Initially the blast furnace is charged with sintered ore, limestone (to produce slag), and coke (preheated to 800 °C). The air blast is also preheated to 900 °C.

Electrolytic zinc

The roasted concentrate from the crude sulphide ore is leached with sulphuric acid (i.e. the zinc oxide is turned into soluble sulphate) and unwanted metal impurities are removed by a succession of chemical treatments. The zinc sulphate/sulphuric acid mixture is then electrolysed, using lead anodes and aluminium cathodes. Lead is unaffected by the fairly dilute acid and aluminium is chosen for the cathodes because it is a good conductor and the zinc is easily stripped from it (mechanically) when the deposited layer of zinc has become thick enough. The anodes and cathodes are very close together and the electrolyte circulates continuously between them. The cells operate at 4 V with a very high current density and the zinc stripped from the cathodes is 99.9% pure.

At the anodes, hydroxide ions are preferentially discharged from the aqueous solution and oxygen gas is evolved.

$$2OH^- - 2e^- \longrightarrow H_2O + O$$
$$2O \longrightarrow O_2$$

At the cathode, zinc metal is deposited

$$Zn^{2+} + 2e^- \longrightarrow Zn$$

Zinc has a relative density of 7.1 at 20 °C and has chemical properties that are generally predictable from its position in the Periodic Classification and the activity series. It has good resistance to atmospheric corrosion. Granulated zinc reacts readily with dilute acids. Zinc, like aluminium, is amphoteric in nature and will react with a strong alkali, such as sodium hydroxide, to give zincate ions with the evolution of hydrogen:

$$Zn + 2OH^- \longrightarrow ZnO_2^{2-} + H_2$$

Zinc is an important reducing agent and the metal in powder form is much used for this purpose in organic chemistry. Many compounds derived from hydrocarbons can be reduced to hydrocarbons again by heating with zinc dust.

The chief uses of zinc are the manufacture of 'galvanized' iron, preparation of alloys such as brass, for alloy die-castings (e.g. petrol pumps in car engines, carburettors, etc.), as a building material for roofs, flashings, etc., and for the preparation of some pigments.

'Galvanized' iron is iron coated with zinc which acts as a 'sacrificial' metal to protect the iron. If the coating is broken and corrosion begins, it is the zinc that disappears and not the iron because, unlike tin, zinc comes above iron in the activity series. Zinc is poisonous, so the zinc-coated iron is never used for foodstuffs but only for domestic ware such as baths, buckets, etc. The cheapest way of applying the zinc is by the 'hot-dip' process. The iron is dipped into molten zinc and withdrawn. A quick withdrawal gives a thick coating; a slow withdrawal gives a thinner one, because the molten zinc can drain off before it finally solidifies. More refined ways of coating the iron are by spraying molten zinc onto the iron, by 'sherardizing' (heating small iron objects in contact with zinc dust), and by electrolysis. The last method is expensive and is only used for small objects that need a particularly good finish. For die-casting alloys the zinc may be alloyed with small amounts of magnesium and aluminium. As pigments, zinc oxide ('zinc white') and zinc chromate (yellow) are used.

Zinc oxide, ZnO

This white oxide has the unusual property of turning yellow when heated and returning to white again when cooled. It is not a chemical change but is due to a change in its structure. Zinc oxide is conveniently obtained in the laboratory by the pyrolysis of hydroxide, carbonate or nitrate. It is amphoteric, reacting with acids to give the corresponding salts and caustic alkalis to give zincates. As a pigment, although it is poisonous, it is not blackened by hydrogen sulphide because zinc sulphide is also white. Zinc oxide is a mild antiseptic and is used in pharmacy as a 'zinc ointment' preparation.

Zinc hydroxide, $Zn(OH)_2$

This forms as a white flocculent (i.e. looking like wool) precipitate when an alkali-metal hydroxide solution, or ammonia solution, is added to the solution of a zinc salt. The precipitate will dissolve in excess of any of these reagents. With alkali-metal hydroxides zincate ions are formed and with ammonia complex ions of formula

$$[Zn(NH_3)_4]^{2+} \qquad \text{(Section 37).}$$

Zinc carbonate, $ZnCO_3$

This insoluble carbonate can only be precipitated in pure form by using sodium (or potassium) hydrogencarbonate solution and a solution of a zinc salt. Normal alkali metal carbonates precipitate a basic carbonate (Section 37).

Zinc chloride, $ZnCl_2$

Like magnesium chloride, this is deliquescent and is partially hydrolysed when heated with water. The anhydrous salt can be obtained by direct combination of its elements. Concentrated solutions of zinc chloride can be corrosive on the skin or other organic materials. A solution of this salt is used in some soldering processes for the removal of oxide film from the metal to be soldered.

Zinc sulphate, $ZnSO_4, 7H_2O$

'White vitriol' (Section 69). This hydrated zinc sulphate when heated loses six molecules of water at 100 °C and the remaining one at 450 °C. (Compare with the action of heat on hydrated copper(II) sulphate 'blue vitriol' (Section 29).)

46   Energy Changes in Chemical Reactions

All objects around us are made up of vast numbers of atoms arranged in different ways. They are all forms of **matter** and all matter has **mass.** Everything that is observed, however, cannot be explained in terms of matter. Matter can be in motion, as for example the second hand of a watch, but it would be hard to imagine that there are atoms of movement. Movement is quite different in kind from matter. In the seventeenth century it was believed that particles of heat were thrown out by a fire, but heat, like movement, is different from matter. They are both types of **energy.** There are many kinds of energy including kinetic (energy of movement), heat, light, electrical, chemical, and nuclear (or atomic). They are all related to one another because any form of energy can be converted into any other form of energy. For example, electrical energy can be converted into light (with some heat energy) by passing an electric current through a light bulb. Just as it is impossible to create or destroy matter in a chemical reaction, so it is impossible to create or destroy energy in any process apart from a nuclear one. With this one exception, *the total energy before a change must equal the total energy after the change.* This is the *Law of Conservation of Energy.*

Energy and matter

When a chemical reaction occurs, energy in some form is either absorbed or evolved. From the Law of Convervation of Energy it follows that there must be a corresponding change in the chemical energy (or 'internal energy') of the substances involved. This chemical energy may be considered to be stored in the bonds linking atoms or ions.

Processes in which **energy is given out** are known as **exothermic** processes.

In these cases, the chemical energy of the products is less than that of the original substances; the difference is given to the surroundings. A familiar laboratory example is the burning of magnesium ribbon in air. Much energy is evolved in this reaction – some of it as heat and some as light. In the simple cell used in a torch battery, chemical changes occur in which energy is given out. Some of this may be obtained as electrical energy although some also appears as heat. When a bullet is fired, the chemical energy of the explosive material is given off partly as the kinetic energies of the bullet and the recoiling gun, partly as the kinetic energy of the molecules in the air causing sound waves, partly as heat, and partly as light.

Processes in which **energy is absorbed** are known as **endothermic** processes.

In these cases, the products have more energy than the reacting substances and the extra energy is absorbed from the surroundings. When compounds are decomposed by electrolysis (Section 21) they absorb energy in the form of electrical energy. In some reactions, energy can be absorbed as light. The light passing through the lens of a camera provides energy to decompose some of the silver bromide on the film – the basis of modern photography.

Energy changes in everyday life

FUELS: Fuels are solids, liquids or gases which will burn in air to produce energy which man can harness for his own purposes. Fuels always produce heat but, sometimes, other forms of energy are produced as well, e.g. the kinetic energy given to a car or lorry when petrol or diesel oil is burnt in the engine. Much research has recently been devoted to the development of *fuel cells* which would convert chemical energy from the fuel directly into electrical energy.

At present, most of the energy required by modern industrial civilizations is obtained from the so-called fossil fuels: coal, oil, and natural gas. Coal is thought to be the remains of forests that existed millions of years ago. The trees died and eventually became buried and compressed (fossilized), undergoing some decomposition at the same time. Oil and natural gas were probably formed in a similar way, but starting from living matter in prehistoric seas. Eventually, all fossil fuels will be exhausted but it is to be hoped that other sources will be available before that time.

Coal and oil are complicated mixtures of compounds and are not usually burnt in the crude state in which they are obtained from the ground. Refining processes are used to produce coke and other similar solid fuels from coal. Similarly crude oil can be made to yield a multitude of products (Section 53). The recent discoveries of oil and natural gas under the sea near the coasts of Britain are most important for our economy.

When buying fuels, the industrialist must know how much energy can be obtained from them. It is thus common to refer to the 'calorific value' of a fuel – which is a measure of how much heat is obtained when a certain quantity of the fuel is completely burnt in a plentiful supply of air. Calorific values are sometimes given in rather quaint units that are no longer used in scientific work.

FOOD: All living organisms need energy in order to grow and reproduce. In the case of those that can move – mainly animals – energy is used whenever movement takes place. In the most highly developed animals – the mammals – correct functioning of their complex bodies depends on maintaining the temperature well above normal atmospheric temperature. Warm-blooded animals, including man, use a great deal of energy, as heat, to maintain body temperature. This heat comes from energy stored as chemical energy in complicated molecules. When these molecules are broken down into simple ones by reactions occurring in the body, kinetic energy and heat are released. Mammals obtain the complex molecules by eating food. The important energy-giving foods contain three main types of organic compound: carbohydrates (Section 58), fats, and proteins (Section 60). Mammals breathe in oxygen from the air and use this to convert food molecules into carbon dioxide (which is exhaled from the lungs) and water (which is lost as sweat and urine). A typical reaction is the oxidation of glucose, $C_6H_{12}O_6$:

$$C_6H_{12}O_6 + 6O_2 \longrightarrow 6CO_2 + 6H_2O$$

As much the same thing happens when coal is burnt to produce energy, food is regarded as fuel for the body. Energy values for foods are usually given on diet charts in units known as Calories, a Calorie being defined as the energy (in the form of heat) required to raise the temperature of 1 kg of water by 1 °C. (A calorie spelt with a small 'c' is one-thousandth of a Calorie.) Cheese, for example, may be quoted on a diet chart as having a value of 120 Cal/oz. This means that the body can convert 1 oz (28 g) of cheese into carbon dioxide and water with the release of 120 Cal or, to use the modern scientific energy unit, 504 kJ of energy in one form or another. Any surplus food, not needed immediately for energy, is stored until required – usually as fat in the tissues of the body. Persistent overeating increases the store of fat and this can throw an unnecessary strain on the heart.

The complex molecules in the bodies of animals are all derived, directly or indirectly, from plants; a cow, for example, derives them from the grass that she eats. Only plants, containing the green colouring matter, chlorophyll, are able to build complex molecules from simple ones such as carbon dioxide and water.

PHOTOSYNTHESIS: Green plants take in simple molecules such as water (through their roots) and carbon dioxide (through pores known as 'stomata' on the underside of their leaves) and build these into complex molecules such as carbohydrates. Complex molecules contain more chemical energy than the simple ones from which they have been made and the plant obtains this extra energy from the Sun. The meaning of 'photosynthesis' should now be clear; the prefix 'photo-' means 'light' and 'synthesis' means 'building up'. A typical reaction is:

$$6CO_2 + 6H_2O \longrightarrow C_6H_{12}O_6 + 6O_2$$

This equation is seen to be the reverse of the last one given. The oxygen given out by photosynthesis compensates for that used in burning and breathing and thus keeps the proportion of oxygen in the air fairly constant. This vital process, however, needs a catalyst and this is provided by **chlorophyll.** In its absence, photosynthesis would occur very slowly, if at all. Chlorophyll is a complicated organic compound containing a little magnesium in its structure. It is present in the leaves of all green plants and is itself coloured green. Most non-green plants, such as mushrooms, cannot build up their own food by photosynthesis and are therefore saprophytic or parasitic. (Copper-beech leaves have chlorophyll but other colouring matter as well.) Photosynthesis is the most important of all endothermic processes. Without it, there could be no life as we know it.

HEAT OF REACTION: A sample of any chemical substance has 'internal energy' which is stored in the bonds between atoms or ions and in the weak attractions due to van der Waals forces (Section 3) between molecules. How much energy is associated with the substance will depend upon its composition, its mass, and the arrangement of the particles. Usually it is impossible to measure the internal energy of a substance. When a change in the arrangement of the particles occurs, as when a chemical reaction takes place, the internal energy of the products will generally be different from that of the original substance(s). The difference in energy is either taken from, or given to, the surroundings. Thus, although the individual internal energies often cannot be measured, the *change* in internal energies can. By suitably arranging the experiment, the difference in energy can be taken up, or given out, as heat – the quantity of which can usually be measured fairly easily.

The actual heat change during a reaction depends upon a number of factors and therefore it is necessary to specify the states of the reactants and products and also the temperature at which the reactants start and the products finish. This temperature is usually chosen to be 298 K (25 °C).

The heat of reaction is the heat evolved or absorbed when a reaction occurs in the amounts (i.e. numbers of moles) indicated by the equation, which must be given.

The heat change is represented by $\triangle H$. If heat is given out, i.e. lost by the reactants when they form products, then $\triangle H$ is negative. In an endothermic reaction, where heat is gained by the reactants to form products, $\triangle H$ is positive. Since such heat changes are usually large, it is convenient to measure them in kJ rather than J.

The burning of hydrogen in oxygen may be represented by the thermo-chemical equation:

$$H_2(g) + \tfrac{1}{2}O_2(g) \longrightarrow H_2O(l) \qquad \triangle H = -287\,kJ$$

(Since the formulae represent moles rather than molecules, fractions are allowable.)

The equation tells us that, at some given temperature (usually 298 K), 1 mole of hydrogen gas will react with $\frac{1}{2}$ mole of oxygen gas to form 1 mole of liquid water, 287 kJ of heat being evolved. In order to decompose a mole of water into hydrogen and oxygen (e.g. by electrolysis) it would be necessary to give it 287 kJ of energy. This would be represented as:

$$H_2O(l) \longrightarrow H_2(g) + \tfrac{1}{2}O_2(g) \qquad \triangle H = +287\,kJ$$

Certain types of reaction are of particular importance in considering heat changes and this has led to special definitions for these cases. The most common are:

(1) **Heat of combustion:** *The heat evolved when 1 mole of a substance is completely burned in oxygen to form products at a given temperature.* The heat of combustion of hydrogen is $-287\,kJ\,mol^{-1}$.

(2) **Heat of formation:** *The heat evolved, or absorbed, when 1 mole of a compound is formed from its elements in their usual states at the temperature concerned.* The heat of formation of water is $-287\,kJ\,mol^{-1}$.

Compounds with a negative heat of formation are called endothermic compounds. These are rare (e.g. ethene, ethyne, carbon disulphide) but most explosives are endothermic. Nearly all common laboratory compounds are exothermic.

(3) **Heat of decomposition:** Same definition as in (2) but replace the words 'formed from' with 'decomposed into'. The numerical value is the same as the heat of formation with the sign changed. Thus the heat of decomposition of water is $+287\,kJ\,mol^{-1}$.

(4) **Heat of neutralization:** *The heat evolved when an acid neutralizes a base in such amounts that 1 mole of water is formed.* (Section 35)

(5) Heat of precipitation: *The heat evolved, or absorbed, when reactants are mixed in such amounts that 1 mole of the precipitate is formed.*

Heats of neutralization and precipitation give convincing indirect evidence for the existence of ions.

Experiments show that the heat of neutralization of any strong acid by any strong base is the same per mole of water formed. To take two examples:

(a) $\quad$ $HCl(aq) + NaOH(aq) \longrightarrow NaCl(aq) + H_2O(l) \quad \triangle H = -57.5\,kJ$
(b) $\quad$ $HNO_3(aq) + KOH(aq) \longrightarrow KNO_3(aq) + H_2O(l) \quad \triangle H = -57.5\,kJ$

This appears to be a strange coincidence since the reactants and one of the products are different in each case. Since the same figure is obtained whatever strong acid/strong base pair is used, it must be more than a coincidence. If we assume that strong acids, strong bases, and their salts are completely ionized, however, equation (a) becomes:

$$H^+(aq) + Cl^-(aq) + Na^+(aq) + OH^-(aq) \longrightarrow Na^+(aq) + Cl^-(aq) + H_2O(l)$$
$$\triangle H = -57.5\,kJ$$

which, by ignoring 'spectator' ions, reduces to:

$$H^+(aq) + OH^-(aq) \longrightarrow H_2O(l) \quad \triangle H = -57.5\,kJ$$

(*Note:* This reaction must not be confused with the formation of water from its *elements.*)

Equation (b) reduces to exactly the same equation. Thus, on the assumption that the reactants and salt product are completely ionized, the identical values for the heat of neutralization are explained.

Considering precipitation reactions, it is found that the heat of precipitation of, say, lead iodide is the same whether 1 mole of lead iodide is obtained by mixing solutions containing suitable amounts of lead nitrate and potassium iodide, or of lead ethanoate (acetate) and sodium iodide. In each case, because the salts are ionized, the reaction is the same:

$$Pb^{2+}(aq) + 2I^-(aq) \longrightarrow PbI_2(s)$$

47 Reaction Rate. Catalysis

REACTION RATE

Importance of reaction rate

The time that a reaction takes is of great importance to the industrial chemist. If he can arrange for a substance to be produced rapidly, he can make more of it in a given time and so lessen costs. He must know, therefore, how various conditions can affect the speed of a reaction.

Theoretical chemists also have an interest in the ways in which reaction rate varies with conditions. By studying these variations, they can make important discoveries about the way in which two substances react and what other substances are formed on the way to the final products.

Measurement of reaction rate

Reaction rate is a measure of how quickly reactants are disappearing or how quickly products are appearing. The actual quantities of reactants and products will depend upon how much material the experimenter starts with and will vary from one experiment to another. To overcome this, the *concentrations* of the materials are considered instead of the quantities.

If a reaction in solution is being studied, the rate of reaction may be measured as a change of concentration per unit of time. It is expressed in moles per litre per second (mol 1^{-1} s^{-1}). For a gas reaction, it is convenient to consider the change in pressure per unit of time instead.

In many reactions there is no visible change. Even when a change can clearly be seen it may not be possible to measure the concentrations accurately and quickly with simple apparatus.

Conditions affecting reaction rate

There are three important variable factors that can influence the rate of a reaction:

(1) *How close the reacting particles are to one another.* This factor may be varied by altering the concentrations or, if the reactants are gases, the pressures of the reactants.

(2) *The level of energy supplied to the reaction.* This may be varied by altering the temperature at which the reaction is carried out or, in some cases, the intensity of light that is falling on the reacting mixture.

(3) *Whether a catalyst is present or not.* (See later in this section.)

EFFECT OF CONCENTRATION: If two molecules are to react, they must be in the same place at the same time. Therefore, in order to react, two molecules must collide or, at least, come very close to each other. The rate at which molecules react will depend upon the rate at which they collide. Often very few collisions result in reaction but, in a given case, the number of collisions which result in reaction is always a constant fraction of the total number of collisions, provided that the temperature remains constant. Thus it follows that the greater the number of collisions, the more rapid the reaction. But, clearly, the greater the number of reactant molecules there are in a given space, the greater will be the number of collisions between them. Therefore, as the concentration of a reactant increases, so does the rate of reaction.

EFFECT OF PRESSURE: This is only important when gases are reacting. The greater the number of gas molecules in a given space, the greater will be the number of collisions in a given time and so the faster will be the reaction rate. A convenient measure of the number of gas molecules in a given space is the pressure exerted by the gas. Therefore, as the pressure of a gas is increased, so also is the reaction rate.

EFFECT OF TEMPERATURE: When two molecules collide, they will only react together if they have sufficient energy. This energy is called the **activation energy.** The reaction rate will therefore depend upon the proportion of molecules which have this, or a higher, energy.

The average energy of particles in a gas, liquid or solid is measured by the temperature. In any collection of particles, however, some will have a very low, and some a very high, energy. An individual particle will change its energy each time it collides with another particle. At any instant, there is a distribution of energies. Since enormous numbers of particles are involved, the graph showing the distribution of energies has the form of a smooth curve (Figure 23(a)). It will be seen that very few particles have either a very high or a very low energy and that most have an energy in the region of the average.

The proportion of the particles which has an energy equal to, or greater than, the activation energy is represented by the shaded area in Figure 23(b).

If the temperature is raised, the average energy is increased by the heat supplied and the curve becomes flattened. Now a larger proportion of the particles have an energy equal to, or greater than, the activation energy. This is shown in Figure 23(c).

It will be seen that a relatively small increase in average energy (i.e. temperature) has a relatively large effect on the proportion of particles with energy equal to, or greater than, the activation energy. This explains why the effect of temperature on reaction rate is so important.

Increasing the average energy of the molecules will also increase the number of collisions between these molecules per second. This effect will also increase the reaction rate in the same way that increasing the concentration does. However, the increase brought about by this factor is quite small compared with that brought about by increasing the proportion of particles having energy equal to, or greater than, the activation energy.

EFFECT OF LIGHT: Sometimes molecules can absorb energy in the form of light rather than heat. When hydrogen and chlorine react, energy may be absorbed in this way. If a mixture of equal volumes of hydrogen and chlorine is left exposed to dull daylight for some days, it will be found that they have reacted together to form hydrogen chloride and neither of the original gases remains. If a similar mixture is exposed to bright sunlight or the light from burning magnesium ribbon, a dangerous explosion can occur. Such reactions

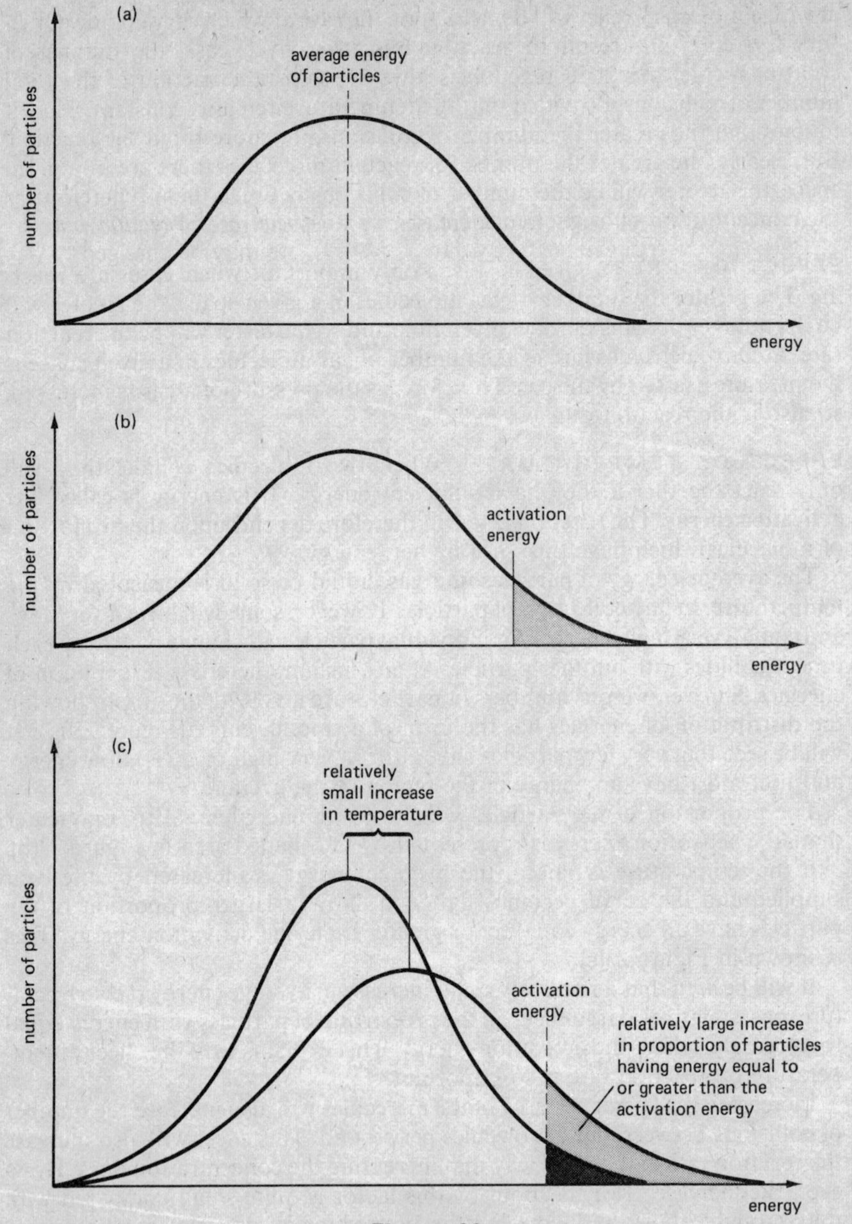

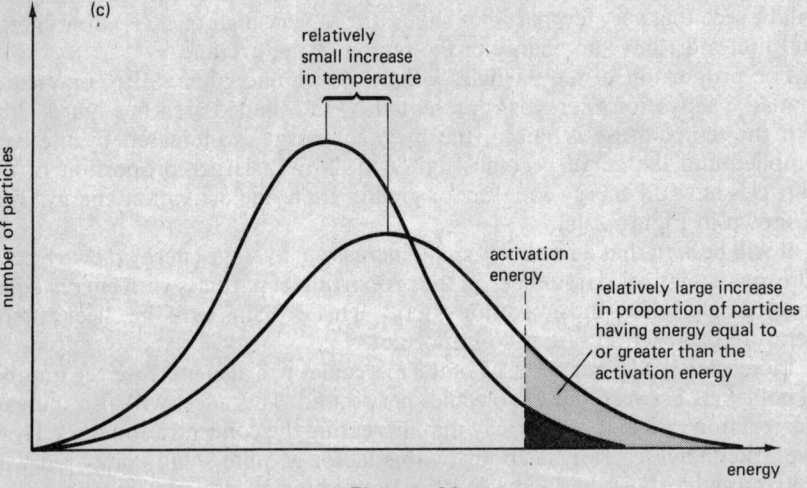

Figure 23

are called **photochemical** reactions to emphasize that energy is absorbed as light.

A more familiar example of a photochemical reaction is the decomposition of silver bromide – the basis of photography. A photographic film consists of a very thin layer of gelatin containing colloidal silver bromide supported on a plastic material. When light falls on the film, it causes decomposition of the silver bromide and the extent of the decomposition depends upon the brightness of the light.

The most important of all photochemical reactions is photosynthesis (Section 46). The energy needed for this reaction comes from sunlight and chlorophyll acts as the catalyst.

CATALYSIS

A catalyst is a substance that can affect the speed of a reaction but is chemically unchanged at the end. Its physical state may be changed.

Most catalysts increase the rate of reaction and are called **positive catalysts.** Some of these may actually start reactions which otherwise would not occur at all. Some other catalysts, **negative catalysts,** may cause reactions to proceed more slowly. Negative catalysts are added to hydrogen peroxide solutions to prevent them from decomposing during storage.

Ostwald in 1888 compared the action of a catalyst in a reaction with the use of oil in a machine and the analogy is reasonably sound (see Table 14).

Table 14

Oil in a machine	Catalyst in a reaction
(1) A small amount of oil is enough	Usually true of a catalyst
(2) Oiling a machine will not start it	Reactions for which a catalyst is used, nearly always proceed slowly without it
(3) Oiling a machine affects its speed but not the products	Quite relevant to catalysis
(4) Oils are specific. Different machines need different types of oil	Also relevant
(5) Grit or other impurities make the oil useless	Catalysts can be 'poisoned' by impurities such as arsenic or cyanide ion

If the catalyst and the reactants are all in the same physical state, the catalysis is known as **homogeneous catalysis,** e.g. water vapour catalysing the reaction between hydrogen sulphide and sulphur dioxide. A solid catalysing reaction between gases would be an example of **heterogeneous catalysis.** Examples: Haber process (Section 63), Ostwald process (Section 64), Contact process (Section 69). Enzymes are important organic catalysts (Section 58–60).

Effect of catalyst on reaction rate

A positive catalyst increases reaction rate by lowering activation energy. This means that a greater number of the particles of the reactants will have sufficient energy to react when they collide and thus the reaction rate will increase. There are two ways in which catalysts can do this:

(1) CATALYSIS AT SURFACES: The catalyst may adsorb reactant molecules in such a way that they are held together for a much longer period of time than they would be in a random collision. The heat of reaction may also be dissipated.

This is particularly likely to happen at the surface of transition metals (Section 70) or activated carbon (Section 49).

(2) INTERMEDIATE COMPOUND FORMATION: The catalyst combines with one of the reactants, the compound formed reacts with the other, then the intermediate compound breaks down to give products and the original catalyst back again.

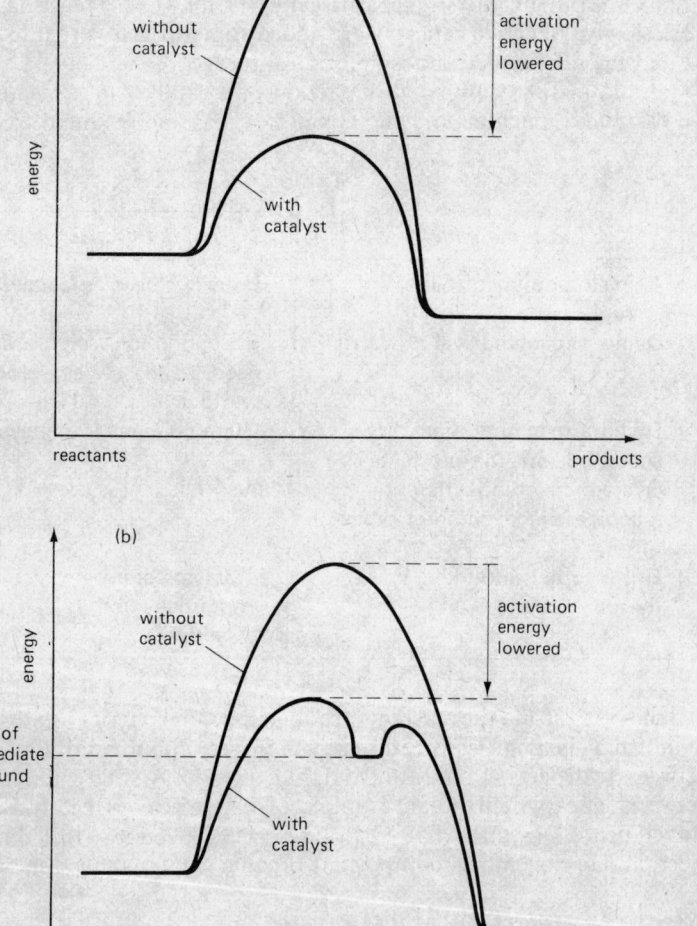

Figure 24   (a) Adsorption by a catalyst.   (b) Formation of an intermediate compound

This is known to happen, for example, when manganese(IV) oxide catalyses the decomposition of potassium chlorate. There can be no doubt at all that in many catalytic reactions, the catalyst does take part in the reaction but is eventually left chemically unchanged.

In either of the cases (1) or (2) above, the energy barrier between reactants and products is lowered as illustrated in Figure 24(a) and (b).

If the original energy barrier is very high, it may be that the reaction cannot occur at all. The effect of using a catalyst then may be to lower the energy barrier sufficiently for a reaction to become possible.

48   Reversible Reaction

If steam is passed over red-hot iron, hydrogen is produced and a black oxide of iron, tri-iron tetroxide, is left in the tube:

$$3Fe(s) + 4H_2O(g) \longrightarrow Fe_3O_4(s) + 4H_2(g)$$

If, on the other hand, hydrogen is passed over heated tri-iron tetroxide, the iron oxide is reduced to iron and steam is formed:

$$Fe_3O_4(s) + 4H_2(g) \longrightarrow 3Fe(s) + 4H_2O(g)$$

This is an example of a **reversible reaction** – one which can proceed in either direction, depending upon the conditions. It is usual to indicate the reversibility in the following way:

$$3Fe(s) + 4H_2O(g) \rightleftharpoons Fe_3O_4(s) + 4H_2(g)$$

Strictly speaking, one cannot refer to 'reactants' and 'products' in a reversible reaction. If an equation is given, however, it is convenient to refer to the substances on the left-hand side as the 'reactants' and to those on the right-hand side as 'products'. Inverted commas are used in this section to indicate this special usage. Similarly, the 'forward' reaction is the reaction reading from left to right and the 'backward' reaction that reading from right to left.

Chemical equilibrium

If iron and steam are placed in a *closed* vessel and heated, they will start to react. As time goes on, the reaction rate will decrease as the concentrations of the reactants decrease. Conversely, the rate of reaction between hydrogen and tri-iron tetroxide will initially be zero (since none of these substances is present). As the concentrations of hydrogen and tri-iron tetroxide increase, however, the reaction rate will increase. There must come a time when the 'backward' reaction is going as fast as the 'forward' reaction. At this stage, although the individual atoms and molecules are still reacting, there will be no change in the concentrations of the four substances. Thus *chemical equilibrium is a dynamic equilibrium.*

It is sometimes convenient to regard all chemical reactions as reversible. Reactions, which are normally regarded as going in one direction only, can be thought of as equilibrium reactions in which the equilibrium concentrations of the 'products' are very high compared with those of the 'reactants'.

Changing the equilibrium concentrations

A more important reversible reaction than the one between iron and steam is that between nitrogen and hydrogen:

$$N_2(g) + 3H_2(g) \rightleftharpoons 2NH_3(g)$$

This reaction is the basis of the Haber process for the synthesis of ammonia (Section 63). To make ammonia as economically as possible, it is important that the equilibrium concentration of ammonia should be increased to the maximum.

It is found that equilibrium concentrations can be altered by most of the factors that affect reaction rates (Section 47). The way in which these factors change equilibrium concentrations may be predicted by using an important principle stated by *Le Chatelier* in 1888. This principle (sometimes called the 'Law of Mobile Equilibrium') states that:

If a stress is applied to a system—either chemical or physical—that is in equilibrium, the system will adjust itself so as to minimize the effect of the stress.

It is a principle of wide application, but here it will be applied to considering the effect of changing conditions upon chemical reactions that are in dynamic equilibrium.

Effect of concentration on equilibrium

If a very dilute solution of an iron(III) salt (e.g. iron(III) chloride) is mixed with a very dilute solution of a thiocyanate (e.g. potassium thiocyanate), a red colour is produced by a reversible reaction represented by the equation:

$$Fe^{3+}(aq) + 6SCN^-(aq) \rightleftharpoons [Fe(SCN)_6]^{3-}(aq)$$

The red colour is due to the $[Fe(SCN)_6]^{3-}$ complex ions. If a further quantity of iron(III) salt is added to some of the red solution, the deepness of the red colour increases, showing that more of the complex ions are formed. Similarly, if more thiocyanate is added, the red colour again increases in intensity. Increasing the concentrations of either of the 'reactants' moves the equilibrium to the right.

This is just what might be expected from Le Chatelier's Principle. By increasing the concentration of iron(III) ions, the system is having a stress applied to it. It will react in such a way as to reduce the concentration of iron(III) ions again. This can only be done by converting some of the iron(III) ions into the complex ions. Exactly similar reasoning applies to the case of adding more thiocyanate ions.

Formation of gases and precipitates

If concentrated sulphuric acid is mixed with a chloride (e.g. sodium chloride) in a closed vessel, a reversible reaction occurs and an equilibrium mixture is obtained:

$$H_2SO_4 + Cl^- \rightleftharpoons HSO_4^- + HCl$$

If the reaction is performed in a vessel open to the air, the hydrogen chloride gas escapes. This means that the concentration of hydrogen chloride in contact with hydrogensulphate ions is very small. Therefore, the 'backward' reaction cannot occur to any great extent. The result is that the 'reactants' are completely converted into 'products'.

When hydrogen sulphide reacts with iron(III) ions in solution, a reversible reaction would be expected:

$$2Fe^{3+}(aq) + H_2S(g) \rightleftharpoons 2Fe^{2+}(aq) + 2H^+(aq) + S(s)$$

As the sulphur precipitates, however, it is almost completely removed from the presence of iron(III) ions and so the 'backward' reaction occurs at only a very slow rate compared with that of the 'forward' reaction. A complete conversion of 'reactants' to 'products' takes place.

Any reaction in which a gas is evolved and allowed to escape, or in which a precipitate is formed, will proceed in one direction only; even though a reversible reaction might be expected.

Effect of pressure on equilibrium

Pressure will have a marked effect on equilibrium concentrations only in the case of reversible reactions between gases. Consider again the Haber process:

$$N_2(g) + 3H_2(g) \rightleftharpoons 2NH_3(g)$$

When ammonia is formed from its elements, the number of molecules present is halved, since the equation shows that four molecules of 'reactants' produce only two molecules of 'product'. Suppose that a mixture of the three gases is in equilibrium. The pressure exerted by the molecules will be a measure of the total number of molecules present in the vessel.

If the pressure on the mixture is increased, the position of equilibrium will move in such a way as to reduce the pressure i.e. to reduce the number of molecules present.

Therefore, increasing the pressure will cause more nitrogen to react with hydrogen and form more ammonia because, in this way, the total number of molecules will be reduced. In industry, pressure from 200 up to 1000 atm are used. The higher pressures produce a higher percentage of ammonia but are more dangerous.

Pressure will have no effect on an equilibrium between gases if equal numbers of molecules appear on both sides of the equation.

Effect of temperature on equilibrium

If a jar of nitrogen monoxide is exposed to the air, the colourless gas reacts with the oxygen to form brown fumes of nitrogen dioxide (Section 62). The reaction is exothermic.

If nitrogen dioxide is placed in a tube and corked lightly to allow for expansion, a change is observed on heating strongly. The brown gas becomes lighter in colour as the temperature rises above 140 °C until eventually, at 620 °C, it turns completely colourless. The reaction is the reverse of the one given in the previous paragraph.

The decomposition of nitrogen dioxide on heating is a reversible reaction and the equation may be written:

$$2NO_2 \rightleftharpoons 2NO + O_2 \qquad \triangle H = +113\,kJ$$
$$\text{brown} \quad \text{colourless}$$

The second experiment shows that, as the temperature is raised, the equilibrium concentrations change in such a way as to absorb some of the heat supplied. This is exactly what would be predicted by Le Chatelier's Principle.

Raising the temperature of any system in equilibrium will cause it to react in such a way that heat is absorbed. The converse is also true.

Effect of a catalyst on equilibrium

In the manufacture of sulphuric acid (Section 69), sulphur trioxide is made by oxidizing the dioxide:

$$2SO_2(g) + O_2(g) \rightleftharpoons 2SO_3(g)$$

This is a reversible reaction. Since the system is slow in attaining an equilibrium, a catalyst is used to speed the process. Originally, platinum was used for this purpose, but nowadays vanadium(V) oxide is used instead. However, it was found that the equilibrium concentrations of sulphur dioxide, oxygen, and sulphur trioxide are quite independent of the catalyst used or, indeed, the absence of a catalyst. In this case, the presence of a catalyst has no effect on the equilibrium concentrations, and this is found to be true for all equilibrium reactions.

A catalyst will increase the rate at which a system comes to equilibrium, but it will not alter the relative concentrations at equilibrium.

This is to be expected on theoretical grounds, because a catalyst lowers the energy barrier between 'reactants' and 'products'. If the barrier is lowered for the reaction proceeding in one direction, it is also lowered for the reaction proceeding in the opposite direction. Thus, any catalyst which increases the rate of a 'forward' reaction, will likewise increase the rate of the corresponding 'backward' reaction.

49   Carbon (Group IV) and its Oxides

CARBON

Carbon is a non-metallic element whose compounds provide the vast field of organic chemistry (Sections 52–60). Its atoms have the power, greater than that of any other element, of linking to one another by covalent bonds to form chains and rings (catenation). Its valence is always four. Carbon monoxide, where its valence appears to be two, is an unsaturated compound (see later in this section).

Carbon has two distinct crystalline allotropic forms that differ markedly from one another – graphite and diamond. Its allotropy is monotropy (Section 50). So-called amorphous forms of carbon have mostly proved to have imperfect graphitic structures. It has been shown that if 1 g of each allotrope is burnt in excess of oxygen, 3.67 g of carbon dioxide is formed in each case and no other product.

Table 15

Graphite	Diamond
Dark-grey; opaque	Colourless; transparent; very high refractive index (2.42)
Soft (marks paper)	Hardest substance known
Layered structure	Rigid structure
Conducts electricity	Does not conduct electricity
Density: 2.22 g cm^{-3}	3.51 g cm^{-3}
Burns in oxygen: $c.\ 700\,^{\circ}\text{C}$	$c.\ 900\,^{\circ}\text{C}$
Heat of combustion: -394 kJ mol^{-1}	-395 kJ mol^{-1}

The great differences are explained by the different structure of the two allotropes, see Figure 25. The structure of graphite is that of layers of carbon atoms, joined with one another to form regular hexagons (like a honeycomb) the bonds being covalent ones. Each carbon atom has four valence electrons and the honeycomb structure involves the joining together of 'benzene rings' (Section 54). The layers are held together by weak van der Waals forces (Section 3) and the softness of graphite is due, at least in part, to the ability of one layer to slide over the next when under stress. The carbon atoms of a benzene ring each have one 'spare' electron and this enables graphite to conduct by the movement of these electrons. It has been shown that the conductivity is much greater when a current passes in a direction parallel to the layers than when it is passed at right angles.

Diamond does not conduct electricity because all the electrons are firmly held in the covalent C—C bonds. The carbon atoms are equidistant from one another, arranged tetrahedrally, and form an extremely hard macromolecule ('giant' molecule) of atoms.

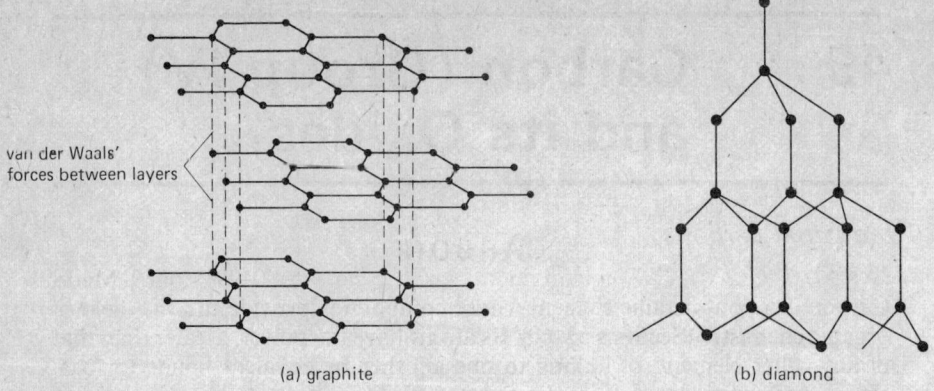

(a) graphite   (b) diamond

Figure 25   Two carbon allotropes

Graphite

Graphite is found in many parts of the world including Sri Lanka, Czechoslovakia, Siberia, Canada, and the U.S.A. (California). Because of its ability to mark paper it has been confused in the past with lead – confusion still perpetuated in the erroneous term 'lead' pencils and 'black lead' polish. The pure element is made by the Acheson Co. (U.S.A.) by heating charcoal, coke or anthracite in an electric furnace with sand (Section 61). Silicon, graphite, and carbon monoxide are formed:

$$SiO_2(s) + 3C(s) \longrightarrow Si(s) + C(s) + 2CO(g)$$
$$\text{graphite}$$

Graphite is an excellent lubricant because the layers can roll up to create miniature roller-bearings. It does not volatilize at high temperatures as most oils do. The Acheson Co. produces colloidal graphite dispersed either over water ('Aquadag') or oil ('Oildag'), 'dag' standing for 'deflocculated Acheson graphite'. These are excellent lubricants. Graphite is not attacked by fused alkali or by chlorine.

Other uses of graphite include preparation of electrodes, 'moderators' for nuclear reactors, and 'lead' pencils. For the latter, powdered graphite is mixed with fine clay. A very soft pencil (6B) contains about 85% graphite: 15% clay. The hardest (9H) might have 70% clay. A recent development has been the manufacture of fine, very-strong graphite fibres. These are prepared by the decomposition of special long-chain polymers, e.g. acrylonitrile (Section 57). The fibres are costly to produce but they are used to reinforce plastics in the CFRP materials ('carbon fibre reinforced plastics').

Diamond

This allotrope is also found native, mainly in the mines of S. Africa and also in Brazil and India. The uses of diamond are well known. For jewellery it is prized for its hardness and brilliance. Its very high refractive index means that light incident at any angle greater than $24\frac{1}{2}°$ will be totally reflected (the corresponding angle for glass being $40\frac{1}{2}°$). Diamond crystals are cut so that all the light incident on any surface is totally reflected and re-emitted. Small imperfect diamonds are useful for cutting tools. These are either fragments rejected by the jeweller or are produced artificially. In 1959 artificial diamonds first became available for industry. To convert ordinary forms of carbon into diamond requires very high pressures and temperatures as well as a catalyst. At a factory in the Irish Republic, carbon is dissolved in molten nickel or cobalt at 1500–2500 °C at 50–100×10^3 atm. pressure. The metal (transition) acts as the catalyst.

Industrial carbons

CARBON BLACK: Finely divided graphitic carbon 90–99% pure. Made either by thermal decomposition of hydrocarbons out of contact with air or by burning oil in the right proportions with air at about 1600 °C. Used for hardening rubber (motor tyres) and for making Indian ink, printer's ink, carbon paper, typewriter ribbon. Also used with coke for making carbon brushes for electrical machines, carbon rods for dry cells, and shoe polish.

CHARCOAL: Prepared by destructive distillation (i.e. heating strongly out of contact with air) of hard wood from deciduous trees. Not so important as formerly but still needed for gunpowder, adsorption of gases or colouring matter (usually in 'activated' form), pharmaceutical preparations, and artists' materials. The 'activated' variety is specially prepared, from wood, peat, coal or coconut shell, to be highly and rapidly adsorbent. The surface area of a porous piece of charcoal is very large but usually the pores are clogged with foreign matter. To make 'activated' carbon, the charcoal is given steam or chemical treatment to cleanse the pores. Gas masks in the two World Wars used 'activated' carbon, but it will only adsorb gases that are fairly easily liquefied such as ammonia, chlorine, and sulphur dioxide. It will *not* adsorb hydrogen, oxygen, nitrogen or carbon monoxide. Well-prepared charcoal can adsorb up to 200 times its own volume of gas. Activated carbons are used for the purification of air or water, for decolourizing solutions, solvent recovery, and as a catalyst or supporter of a catalyst.

'Animal' or 'bone' charcoal is obtained by the destructive distillation of bones and consists mainly of calcium orthophosphate. It is still used for decolourizing natural sugar.

COAL: Has been formed by the decay of old wood over hundreds of years and probably through the intermediate stage of peat. Peat compressed by earth movements eventually gives coal. Carbon contents: wood 50–55%; peat 55–60%; lignite 60–75%; bituminous coal 75–90%; anthracite over 90%. As the product ages, the proportion of carbon rises as the proportions of oxygen and volatile matter fall. Bituminous coal burns with a bright, smoky flame whereas anthracite burns with little flame and leaves hardly any ash. The structure of coal is now fairly well known. It is extremely complicated, with the carbon mainly in ring-structures and oxygen appearing here and there as hydroxyl, $-OH$, or carbonyl, $=CO$, groups.

COKE FROM COAL: Made by the destructive distillation of coal (85–90% carbon) in vertical retorts constituting what is known as a 'coke-oven' plant. Formerly an important by-product of the manufacture of town gas. Still made for use in blast furnaces and for the preparation of gaseous fuels, though these are being ousted by oil fuels and natural gas. Coke ovens operate at temperatures up to 1000 °C and all volatile and tarry matter is expelled. If the temperature is kept down to about 600 °C, the process is known as 'low-temperature carbonization'. It produces fuels such as 'coalite' which still contain the less-volatile tarry matter.

COKE FROM PETROLEUM: Has the advantages that it is purer than that obtained from coal and has a more regular structure. Minimum content of carbon 98.5%. Prepared from the residues of petroleum distillation and used for the manufacture of electrodes for the aluminium and steel industries, brushes, bearings, crucibles, and linings for chemical plant.

SUGAR-CARBON (SUGAR-CHAR): A pure form of carbon made by the dehydration of sucrose, either by pyrolysis or by treatment with concentrated sulphuric acid. An expensive way of obtaining pure carbon and has no industrial use.

SOOT: This can be deposited in chimneys of domestic or industrial heating plant. The Clean Air Act has stopped it from polluting the atmosphere and industrialists now realize that black smoke advertises waste of fuel. Soot from coal always contains a small amount of ammonium compounds and has a slight value as a fertilizer.

Carbon can combine with various metals forming carbides and one of the most important of these is tungsten carbide, WC, which is almost as hard as diamond. The importance of carbon as a reducing agent appears in other sections, e.g. the metallurgies of lead (Section 44), zinc (Section 45), and iron (Section 71).

CARBON DIOXIDE, CO_2

When carbon burns in excess air or oxygen, the colourless, odourless gas, carbon dioxide is formed:

$$C + O_2 \longrightarrow CO_2$$

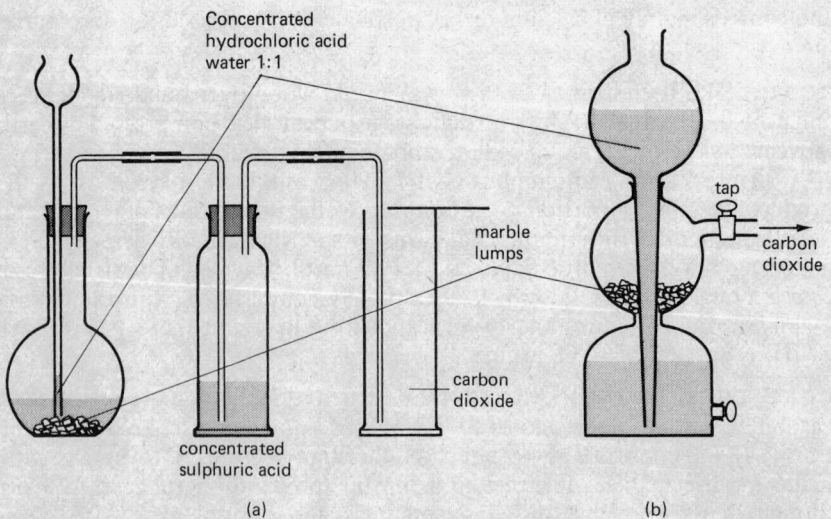

Figure 26 Preparation of carbon dioxide: (a) apparatus for preparing dry carbon dioxide and (b) Kipp's apparatus for the laboratory preparation of the gas

In the laboratory, it is usually prepared by the action of hydrochloric acid on marble (Section 75). A Kipp's apparatus is convenient (Figure 26). It is particularly useful when the gas is needed intermittently and in fairly small quantities. When the tap is opened, acid falls from the top chamber into the bottom one. The bottom compartment steadily fills, the acid rising eventually to the middle chamber where it reacts with the marble to produce the gas. When the tap is closed, the gas is confined and its pressure soon forces the acid from the marble, some of the acid being returned to the top chamber. The apparatus can be used for the preparation of other gases where no heating is required, e.g. hydrogen sulphide (Section 75).

The gas does not support the combustion of feebly-burning substances (hence its use in fire extinguishers (Section 55)). Substances that burn at a high temperature, e.g. magnesium, can decompose the gas and continue to burn:

$$CO_2 + 2Mg \longrightarrow 2MgO + C$$

If burnt in a gas jar the black specks of carbon can be seen.

The usual test for carbon dioxide is its action on lime-water (calcium hydroxide solution). First, a white precipitate of calcium carbonate is formed and this then redissolves as the soluble hydrogencarbonate forms. (See Section 27 'hardness' of water.) The gas is fairly soluble in cold water (1 vol. gas:1 vol. water) under ordinary conditions. Under a few atmospheres pressure it dissolves more readily giving 'soda-water'. The solution is a weak acid, carbonic acid, that turns blue litmus solution a wine-red colour. The solution contains small quantities of oxonium and hydrogencarbonate ions:

$$H_2O + CO_2 \rightleftharpoons H_2CO_3 \overset{H_2O}{\rightleftharpoons} H_3O^+ + HCO_3^-$$

Under normal conditions the pH is 3.7. Carbonic acid is a dibasic acid (Section 33) giving both normal and hydrogencarbonates. All normal carbonates are insoluble in water (except those of the very electropositive elements K, Na, and ammonium) and are obtained by ion aggregation (Section 37). Only sodium and potassium carbonates are not decomposed by heating. In the activity series (Section 6) from lithium onwards, carbonates break down more and more easily into oxide and carbon dioxide, although in the case of Group II metals, e.g. calcium, the change is a dissociation rather than a decomposition. Copper(II) carbonate decomposes easily.

Hydrogencarbonates are not numerous. The only solid ones known are those of the alkali metals (excluding lithium) and ammonium. The hydrogencarbonates of lithium, Group II metals, and iron(II) are known only in aqueous solution. On heating, either dry or in solution, hydrogencarbonate ions split up giving carbonate ions, carbon dioxide, and water:

$$2HCO_3^- \longrightarrow CO_3^{2-} + CO_2 + H_2O$$

Sodium hydrogencarbonate can thus be distinguished from sodium carbonate. As potassium hydrogencarbonate, $KHCO_3$, is more soluble than sodium hydrogen carbonate, potassium hydroxide solution is a more efficient absorbing agent for carbon dioxide than sodium hydroxide solution of the same molarity.

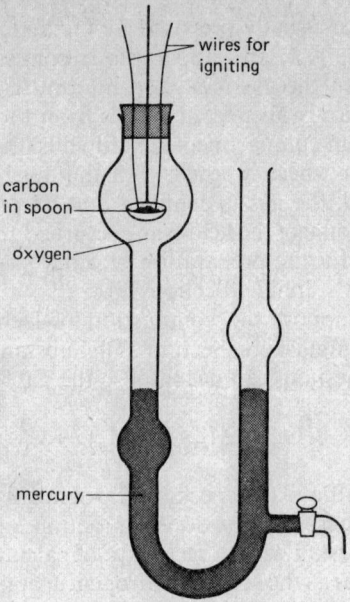

Figure 27

The formula for a molecule of carbon dioxide was established when it was found, using an apparatus similar to the one shown in Figure 27, that pure carbon burns in oxygen with no change in volume when the apparatus cools to its original temperature.

This must mean, from Avogadro's Law, that every molecule of oxygen used has been replaced by a molecule of carbon dioxide. Knowing that the formula for a molecule of oxygen is O_2 (Section 3) it follows that the molecular formula of carbon dioxide is C_xO_2:

$$xC + O_2 \longrightarrow C_xO_2$$

The relative density of carbon dioxide is 22. Therefore, the relative molecular mass is 44 and x must be 1.

The chief uses of carbon dioxide include: refrigeration, making 'fizzy' drinks, in fire extinguishers, for fruit storage, preparation of carbonates and hydrogen-carbonates. The gas is available stored in cylinders under considerable pressure. If the valve of a cylinder is covered with a velvet bag before opening, the sudden expansion cools the gas sufficiently to allow the formation of solid carbon dioxide ('carbon dioxide snow') in the bag.

CARBON MONOXIDE, CO

This gas is formed whenever carbonaceous material burns in an insufficient supply of air (carbon itself may be formed as well). It is colourless, has about the same density as air (14:14.4) and is extremely poisonous. 1% of the monoxide in breathed air may kill.

Chemically, carbon monoxide is an **unsaturated** compound, i.e. all the valence bonds of the carbon atom are not in use. The carbon atom has four valence electrons and only two are used in the linkage with the oxygen atom. Two electrons are not used (compare with tin(II) and lead(II) in Sections 43 and 44). The gas, therefore, readily unites with two other monovalent atoms or one divalent atom to form a saturated compound. With chlorine gas it gives an addition compound; a poisonous gas called carbonyl chloride (phosgene):

$$CO + Cl_2 \longrightarrow COCl_2$$

In air or oxygen, carbon monoxide burns with a blue flame forming the dioxide:

$$2CO + O_2 \longrightarrow 2CO_2$$

Carbon monoxide is a deadly poison because it combines with human red blood corpuscles to form a bright scarlet compound called 'carboxyhaemo-globin' and this reaction is *not* reversible. In normal respiration, oxygen forms a bright-red compound 'oxyhaemoglobin' which readily decomposes to give the necessary oxygen to cells and tissues. Being colourless and odourless and having almost the same density as air, carbon monoxide is particularly danger-ous. It is present in the exhaust gases from motor vehicles and this can be fatal in a confined space. In the open air, because of the relatively large bulk of oxygen in the atmosphere it is soon oxidized to the dioxide – a change that can be catalysed by manganese(IV) oxide or copper(II) oxide.

Because it can bring into use two spare electrons per molecule, carbon monoxide is a powerful reducing agent and when carbon reduces the ores of lead or zinc it is largely the formation of the monoxide that is responsible. It will not, however, reduce the oxide of any very electropositive element, such as aluminium and those elements above it in the activity series.

The molecular formula of the monoxide was established when it was found that two volumes of the gas require just one volume of oxygen to form two volumes of carbon dioxide (all under the same physical conditions):

$$2 \text{ vols gas} + 1 \text{ vol. oxygen} \longrightarrow 2 \text{ vols carbon dioxide}$$

By Avogadro's Law:

$$2 \text{ moles gas} + 1 \text{ mole oxygen} \longrightarrow 2 \text{ moles carbon dioxide}$$

Therefore $\quad 2 C_xO_y + \quad O_2 \quad \longrightarrow 2 CO_2$

x and y must each be 1.

Carbon monoxide is readily absorbed by an ammoniacal solution of copper(I) chloride (Section 72).

The monoxide can be obtained from the dioxide by passing the latter through a long hard-glass tube containing red-hot charcoal (Figure 28). The issuing gas is passed through two wash bottles containing potassium hydroxide solution to remove any unchanged dioxide and is finally collected over water in which it is practically insoluble. The gas collected should have no effect on lime-water. On burning, the flame is described as a blue, lambent ('playing on the surface') flame. After burning, lime-water will be turned turbid. The pure gas can be obtained from methanoic (formic) acid (Section 75).

Carbon monoxide used to be of great importance as a constituent of fuel gases such as producer gas and water gas. Producer gas was obtained by passing air over red-hot carbon:

$$2C + O_2(\text{with } N_2) \longrightarrow 2CO\ (+N_2) \qquad \triangle H = -227\,\text{kJ}$$

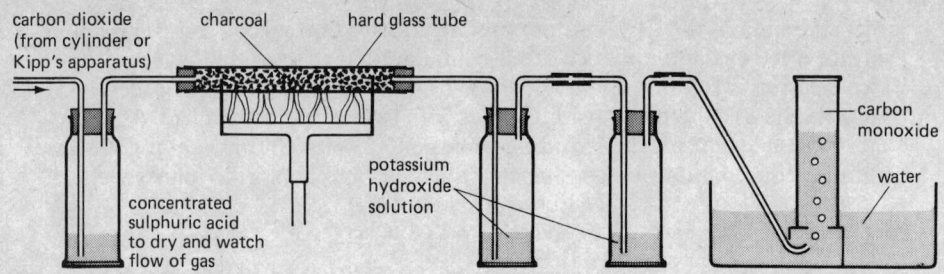

Figure 28 Preparation of carbon dioxide from carbon monoxide

Applying Avogadro's Law, if air is one-fifth oxygen and this volume is doubled, the product should contain one-third carbon monoxide to two-thirds nitrogen by volume. In practice it was never more than 30%.

Water gas was made by passing steam over *white-hot* coke and, theoretically this could produce equal volumes of carbon monoxide and hydrogen:

$$C + H_2O \longrightarrow CO + H_2 \qquad \triangle H = 117\,kJ$$

As this reaction is endothermic, water gas and producer gas were often made alternately to prevent the temperature of the carbon from falling too low; or a mixture of steam and air could be passed to give 'semi-water gas' which contained about 50% nitrogen with roughly equal quantities of hydrogen and carbon monoxide.

All these fuels are poisonous – particularly the water gas – and they have been superseded by gaseous fuels derived from oil or natural gas.

50 Allotropy

A pure element shows **ALLOTROPY** (*from the Greek meaning 'other forms'*) *if it can exist in two or more forms in the same physical state* (*solid, liquid or gas*).

The forms always differ physically and often chemically as well.

Elements that have allotropic forms occur in Groups IV, V, and VI of the Periodic Classification and those considered here include carbon and tin (Group IV), phosphorus (Group V), and oxygen and sulphur (Group VI).

Apart from metals, the bonding between the atoms is always covalent and the difference between the allotropes may be due either to different numbers of atoms in the molecule (where there *are* separate molecules) or to a different arrangement of atoms in the structure. Where the different forms are crystalline, the term **polymorphism** (from the Greek meaning 'many shapes') is sometimes used, but this term could not cover the allotropy of oxygen.

A distinction has been made between two types of allotropy: monotropy and enantiotropy.

Monotropy

Here there is only *one* stable form and the most stable form always has the least intrinsic energy. Thus, for example, it will have the lowest heat of combustion (Section 46). Sometimes a less stable form will show a marked tendency to turn into a more stable one. Examples of monotropy are found in carbon, phosphorus, and oxygen. In monotropy there is no definite transition temperature (see below) between one form and another.

CARBON: Two crystalline forms, graphite and diamond (the former being the more stable). There is no evidence that diamond ever changes into graphite but it has a slightly higher heat of combustion so, theoretically, it is the less stable.

PHOSPHORUS: Two crystalline forms, red and white. The white form does show a tendency to change into the red. The white is cubic with four atoms to the molecule P_4 and the red is rhombohedral (constitution unknown). Another, black, form can exist, but only under very high pressures.

OXYGEN: Oxygen, O_2, and ozone, O_3. Oxygen is the stable form.

Enantiotropy (from the Greek meaning 'opposed forms')

Two or more allotropes of the element can be stable throughout particular ranges of temperature. One form will change into another at a temperature known as the **transition temperature.** Examples are provided by sulphur and tin.

SULPHUR: Two crystalline forms – alpha (rhombic) and beta (monoclinic also sometimes called prismatic). Transition temperature is 95.6 °C. Below this temperature the rhombic form is stable; above it the monoclinic form is stable. Both have the constitution S_8 but there are different arrangements.

TIN: Allotropy is not confined to non-metallic elements. Up to its melting point of 232 °C, tin can have three different crystalline allotropes (the transition temperatures are given above the arrows):

$$\text{Grey tin} \underset{}{\overset{13\,°C}{\rightleftharpoons}} \text{white tin} \underset{}{\overset{161\,°C}{\rightleftharpoons}} \text{rhombic tin}$$
$$\text{(cubic)} \qquad \text{(tetragonal)}$$

Dynamic allotropy

This occurs when two allotropic forms can coexist in equilibrium, e.g. the two liquid allotropes present in molten sulphur. The proportion of each present depends on the temperature.

51   The Carbon Cycle

Vast quantities of carbon dioxide are being poured into the atmosphere every day by the burning of carbonaceous fuels (nearly every common fuel is such) and the respiration of millions of living creatures (Section 30) – all at the expense of oxygen. Yet, even in large towns, the volume percentages of oxygen (21%) and carbon dioxide (0.03%) do not vary much from normal. There must be one or more restoring factors. Priestley, a discoverer of oxygen, found the main one. He investigated what we now call 'photosynthesis' and proved that green plants can, in daylight, take in carbon dioxide from the air, use the carbon in their growth, and *restore* oxygen to the air (Section 46).

Much carbon dioxide is removed from the atmosphere every time it rains and this weakly acid solution filters through the soil turning some insoluble carbonates (e.g. those of calcium, magnesium, and iron(II)) into soluble hydrogencarbonates; but there is no restoration of oxygen here. Some carbon dioxide returns to the air when other acids in the soil (e.g. nitric acid) act on carbonates, but the only oxygen-restoring factor seems to be the process of photosynthesis.

When organic matter decays in contact with air, the carbon returns to the atmosphere as carbon dioxide. If, however, the carbonaceous material rots out of contact with the air, under still water for example, then methane ('marsh gas') (CH_4) is formed.

52   Organic Chemistry

Until the first quarter of the nineteenth century, chemists believed that compounds made by living organisms could not be synthesized in the laboratory. At that time urea, $CO(NH_2)_2$, was a well-known compound obtainable from human urine. When Wöhler, in 1828, obtained urea by accident when he was trying to obtain ammonium thiocyanate, NH_4CNO, from non-living sources, this particular barrier was broken down and now thousands of organic compounds are synthesized, many of them on an industrial scale. (Urea and ammonium thiocyanate have the same atoms in their molecules but they are differently arranged – an example of isomerism (Section 53).) The word 'organic' refers to something living but this is now no longer relevant. The modern definition of *organic chemistry is the chemistry of the carbon compounds.* Compounds such as carbon dioxide, carbonates, etc., which have for so long been associated with inorganic (non-living) material, are still treated in textbooks as being part of inorganic chemistry.

The reason why carbon can be the starting point for such a vast field of chemistry is to be found in its unique power of catenation, that is, the ability of the atoms of carbon to form covalent bonds with each other, more or less indefinitely, to form straight chains, branched chains, and rings of carbon atoms. The first oil well was drilled in Pennsylvania in 1859 and now more than two-thirds of the world's energy needs are supplied by oil and natural gas – all compounds of carbon. A huge petrochemical industry has grown up in the last fifty years and, by 1970, world production (excluding the Communist countries for which no figures are available) was annually over sixty million tonnes of organic chemicals.

Metals do not figure very largely in organic compounds and many thousands of organic compounds involve only a few other elements such as hydrogen, oxygen, nitrogen, the halogens, and sulphur. Industrially, most of them are derived from crude oil (petroleum) and natural gas (methane, CH_4).

53   Saturated Hydrocarbons: Substitution Reactions

A hydrocarbon consists of hydrogen and carbon ONLY and through the power of catenation of the carbon atom (Section 52) there is no limit to the number of hydrocarbons that can be made. Many of them can be obtained from natural petroleum and the simplest of all is methane, CH_4. It is the first of a series of hydrocarbons known as the **alkanes**, formerly called the 'paraffin series' because of their chemical inertness (par-affin means 'little affinity'). They are **saturated hydrocarbons** because the four valence electrons of the carbon atoms are all used to form covalent bonds with carbon or other atoms. If alkanes had been chemically active they would not have remained underground for thousands of years without change.

Methane itself has a simple structure in which four hydrogen atoms are each joined by a covalent bond to the central carbon atom forming a molecule

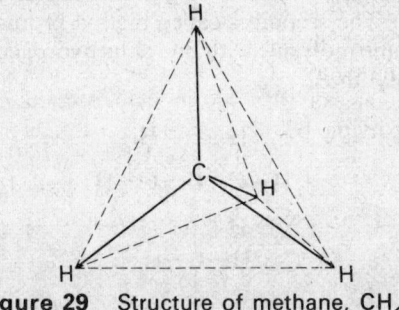

Figure 29   Structure of methane, CH_4

in the shape of a regular tetrahedron (Figure 29). It may be hard to realize that all the hydrocarbon molecules are three dimensional and are not flat as shown on a printed page.

As alkane succeeds alkane, another link is placed in a chain – each link consisting of a carbon atom with two hydrogen atoms attached. Thus the total relative molecular mass increases by 14 at each addition:

$$
\begin{array}{ccc}
\text{H} & \text{H H} & \text{H H H} \\
| & | \ | & | \ | \ | \\
\text{H}-\text{C}-\text{H} & \text{H}-\text{C}-\text{C}-\text{H} & \text{H}-\text{C}-\text{C}-\text{C}-\text{H} \\
| & | \ | & | \ | \ | \\
\text{H} & \text{H H} & \text{H H H}
\end{array}
$$

methane, CH_4 ethane, C_2H_6 propane, C_3H_8

16   30   44

$$
\begin{array}{c}
\text{H H H H} \\
| \ | \ | \ | \\
\text{H}-\text{C}-\text{C}-\text{C}-\text{C}-\text{H} \quad \text{etc.} \\
| \ | \ | \ | \\
\text{H H H H}
\end{array}
$$

butane, C_4H_{10}

58

The general formula of such a series can be expressed as C_nH_{2n+2} where n represents the number of carbon atoms in the molecule. Thus the formula of a hydrocarbon in this series having fifty carbon atoms in its molecule must be $C_{50}H_{102}$.

Although the chemical properties of all members of this series are similar, there is a regular change in the physical properties. As the molecules become larger one would expect the boiling point, for example, to be higher. The first members of the series – as far as butane – are gases. After these, the alkanes are liquids and eventually they are solid (as in paraffin wax). The boiling points of each of the first four alkanes are:

methane, $-164\,°C$ ethane, $-88.6\,°C$ propane, $-44.5\,°C$ butane, $-0.5\,°C$

Pentane C_5H_{12} is a liquid (b.p. $36\,°C$). The names of alkanes have Greek or Latin prefixes from pentane onwards which indicate the number of carbon atoms in the molecule: hexane, heptane, octane, nonane, decane, etc. Such a series is known as a **homologous series,** 'homologous' meaning 'of the same nature or origin'.

From butane onwards there can be more than one compound with the same molecular formula. The molecules have the same number of atoms of each element, but there are different arrangements. The two compounds with the formula C_4H_{10} are butane (with the straight-chain arrangement shown above) and methyl propane. The second is called methyl propane because it can be considered as a propane molecule with one of its hydrogen atoms replaced by a methyl group $(-CH_3)$ thus:

$$
\begin{array}{c}
\text{H H H} \\
| \ | \ | \\
\text{H}-\text{C}-\text{C}-\text{C}-\text{H} \quad \text{methyl propane, } C_4H_{10}\\
| \ \ | \ \ | \\
\text{H} \ \ | \ \ \text{H} \\
\text{H}-\text{C}-\text{H} \\
| \\
\text{H}
\end{array}
$$

Likewise, with C_5H_{12} there are three possibilities. First, there is the straight-chain pentane:

$$H-\underset{\underset{H}{|}}{\overset{\overset{H}{|}}{C}}-\underset{\underset{H}{|}}{\overset{\overset{H}{|}}{C}}-\underset{\underset{H}{|}}{\overset{\overset{H}{|}}{C}}-\underset{\underset{H}{|}}{\overset{\overset{H}{|}}{C}}-\underset{\underset{H}{|}}{\overset{\overset{H}{|}}{C}}-H \quad \text{pentane, } C_5H_{12}$$

Second, there is a compound analogous with methyl propane known as methyl butane:

$$H-\underset{\underset{H}{|}}{\overset{\overset{H}{|}}{C}}-\underset{\underset{H-C-H}{|}}{\overset{\overset{H}{|}}{C}}-\underset{\underset{H}{|}}{\overset{\overset{H}{|}}{C}}-\underset{\underset{H}{|}}{\overset{\overset{H}{|}}{C}}-H \quad \text{methyl butane, } C_5H_{12}$$

Third, there is an arrangement where one carbon atom is centrally placed with four methyl groups spaced round it tetrahedrally. This is dimethyl propane:

$$\text{dimethyl propane, } C_5H_{12}$$

Compounds which have the same molecular formula but differ in the arrangement of atoms in the molecule are known as **isomers** *(from the Greek meaning 'sharing equally').*

It will be realized that, as the number of carbon atoms increases, the possibilities become progressively more numerous. There is now a logical, systematic nomenclature that enables chemists to differentiate between compounds even when their structures are very complicated, but students at an elementary stage would find this more confusing than helpful. Different isomers always have different physical and chemical properties, though sometimes the differences are slight.

Isomerism can occur in many other series besides the alkanes. The earliest example was Wöhler's preparation of urea (Section 52).

Saturated hydrocarbons are obtained from crude oil by a large-scale distillation process. The crude oil is found in various areas including the Middle East, Venezuela, the U.S.A., and the North Sea. Sometimes methane (natural gas) is found with the oil and sometimes the gas alone is discovered. The heavy, unpleasant-smelling dark-green/brown crude oil contains solids and gases in solution and it is first submitted to a primary distillation in a huge fractionating column (Figure 30) which is divided into many stages.

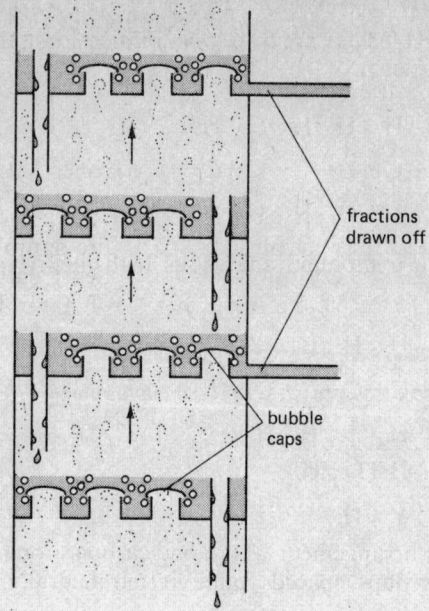

fractions
drawn off

bubble
caps

Figure 30   Part of a fractionating column

Each stage will be operating at a different temperature and the temperature becomes progressively lower as the distance from the heating fuel becomes greater. The vapour from the stage below emerges from under the mushroom-shaped bubble caps. Some of it condenses, but the more volatile compounds pass on to the next stage. When the stage is filled with the liquid as far as the overflow, some returns to the stage below and the process is repeated. Gases are collected from the top of the tower. The top stage will contain the most volatile hydrocarbons and the lower stages those with larger molecules and higher boiling points. Each fraction drawn off, apart perhaps from the most volatile ones, will consist of a complicated mixture of straight-chain hydrocarbons with many branched-chain isomers.

The primary distillation separates the crude oil into gases and three main

Table 16

Fraction	Boiling point range/°C	Number of carbon atoms in the molecules	Description
1. Gases	below 0	1–4	
2. Gasolines	0–65	5–6	
3. Naphtha	65–170	6–10	Contains some carbon-ring compounds (aromatics)
4. Kerosene	170–250	10–14	Complex alkanes and aromatics
5. Gas oil	250–340	14–19	Complex alkanes and aromatics
6. Lubricating oil and wax	340–500	19–35	Complex alkanes and aromatics
7. Bitumen	over 500	over 35	

liquid products: light, middle, and heavy distillates. The second distillation produces the fractions listed in Table 16. (The separation of any one particular hydrocarbon from any fraction would be an extremely involved and costly operation.)

Methane (natural gas), CH_4

Methane has been known as 'marsh gas' and as 'fire damp'–the gas that has caused so many dangerous explosions in coal mines. Coal gas contains from 25–35% methane by volume. Large quantities of natural gas, which have become the starting point of huge chemical industries, have been discovered during the twentieth century in S. France, N. Italy, and the North Sea. It has now become a household word in Britain. It has largely replaced town gas because it is non-toxic, cheap, and has a greater heating-value per unit volume.

In the laboratory, the gas can easily be obtained by heating a mixture of anhydrous sodium ethanoate with soda-lime (Section 75). When ignited in a plentiful supply of air or oxygen it burns, as every hydrocarbon does, to produce carbon dioxide and water vapour. The general equation for the complete combustion of any hydrocarbon (represented here as C_xH_y) must be:

$$C_xH_y + (x + \tfrac{y}{4})O_2 \longrightarrow xCO_2 + \tfrac{y}{2}H_2O$$

For methane, $x = 1$ and $y = 4$ so this gas must need twice its own volume of oxygen for complete combustion or about ten times its own volume of air. This is nearly twice the volume of air that town gas needs, so burners have to be changed when gas fires, cookers, etc. are converted from town to natural gas.

When methane and chlorine are mixed and exposed to diffused sunlight, there is a slow reaction, the hydrogen atoms of the methane molecules being successively replaced by those of chlorine:

(1st stage) $CH_4 + Cl_2 \longrightarrow CH_3Cl + HCl$

Then, in a similar manner CH_2Cl_2, $CHCl_3$, and CCl_4 are formed in turn.

Methane is unable to form addition compounds (Section 54) because its valence electrons are all in use and, apart from its readiness to burn, it is a fairly inert compound. An active electronegative element like chlorine is able, however, to replace the hydrogen atoms and thus form new compounds by **substitution.** There are much better ways of preparing these chloro-compounds and the last two are well known as 'chloroform' $CHCl_3$, and 'carbon tetrachloride', CCl_4. Chemists now prefer to call these four substitution products from methane by their systematic names: chloro-, dichloro-, trichloro-, and tetrachloromethane.

If one hydrogen atom of an alkane molecule is in some way substituted by another atom or group, we obtain a homologous series of compounds with characteristic properties (Table 17).

(The formulae used in Table 16 are described as **condensed structural formulae.** They are sufficient to show which groups are present, without setting out all the individual bonds, and hence save much space.)

It is the hydroxyl group, $-OH$, that gives to alcohols their special properties and it is the carboxyl group, $-COOH$, that gives typical properties

Table 17

Alkane	Chloro-derivative	Alcohol	Acid
CH_4	CH_3Cl	CH_3OH	$H.COOH$ (methanoic)
C_2H_6	C_2H_5Cl	C_2H_5OH	$CH_3.COOH$ (ethanoic)
C_3H_8	C_3H_7Cl	C_3H_7OH	$C_2H_5.COOH$ (propanoic)

to organic acids (Section 56). These, therefore, are known as **functional groups.**
The rest of the molecule contains a **radical** or group of atoms that persists
unchanged through a series of reactions.

    $-CH_3$ is known as the **methyl radical** and is often written 'Me'.

    $-C_2H_5$ is known as the **ethyl radical** and is often written 'Et'.

EtOH represents ethyl alcohol, usually called ethanol because it is the alcohol
derived from ethane. All the derivatives listed above have molecules that consist of a radical linked to a functional group and they are much more
chemically active than the alkanes themselves.

    Methane is a completely covalent compound and it has no polarity. The
introduction of one or more chlorine atoms into the molecule gives it polarity
because the chlorine is so electronegative. Tetrachloromethane, CCl_4, however,
has no polarity because the tetrahedron has a chlorine atom instead of a
hydrogen at each corner and the polarities cancel out.

    Methane has become of great importance to industry because it is now pos-
sible to break its molecules into fragments and from them to obtain unsaturated compounds such as ethene, C_2H_4, and ethyne, C_2H_2, as well as
carbon black (Section 49). Methane can also be converted into hydrogen and
carbon monoxide by passing it with steam over a nickel catalyst at 900 °C:

$$CH_4(g) + H_2O(g) \longrightarrow CO(g) + 3H_2(g)$$

Here two-thirds of the hydrogen comes from the methane and one-third from
the steam.

54 Unsaturated Hydrocarbons: Addition Reactions

Crude petroleum can, by distillation processes, be separated into seven main
fractions (Section 53). There is not, however, an equal demand for all the
products. It is the second fraction – gasoline – that is most required, for this
provides petrol for motor cars. To meet this demand, chemists have devised

means for breaking down the constituents of the higher-boiling fractions in order to produce more motor spirit. The process is known as **cracking** because large molecules break down into smaller ones. At first, high temperatures were employed in what was known as **thermal cracking** but these methods have given way to catalytic processes for bringing about the molecular breakdown. The vapours are passed up through a tower (the **catalytic cracker**) containing a 'fluidized bed' of catalyst at about 500 °C. The catalyst consists of a very finely divided clay material which, when the vapours are passing through it, behaves almost like a liquid. During the molecular breakdown, fine black carbon is formed on the catalyst and this, after a while, makes the catalyst less effective. The carbon has to be burnt off periodically.

When the large molecules crack, unsaturated molecules of ethene, C_2H_4, and propene, C_3H_6, are formed and these are far more reactive than the alkanes. All modern oil refineries now have, often on adjacent sites, enormous chemical plant for making petrochemicals largely derived from unsaturated hydrocarbons produced in this way.

ALKENES

Ethene (ethylene), C_2H_4

Ethene was formerly called 'ethylene' and it was first discovered by Dutch chemists in 1794. Ethene has two hydrogen atoms less per molecule than the corresponding alkane – ethane. This homologous series of unsaturated hydrocarbons is known as the **alkenes** and their general formula is C_nH_{2n}. If there exists an alkene with fifty carbon atoms its formula must be $C_{50}H_{100}$. The first members of the alkene series are:

eth(yl)ene, C_2H_4 prop(yl)ene, C_3H_6 but(yl)ene, C_4H_8

From four carbon atoms per molecule onwards, isomers are possible.

The structure of ethene involves a **double bond** between the two carbon atoms. This is a source of considerable strain in the molecule and hence the double bond breaks easily, making available two unused valence electrons. (The double bond consists of two covalent bonds each involving two electrons.) Thus it is a comparatively easy matter for ethene to form **addition** compounds, the ethene molecule taking up two monovalent atoms or one divalent atom to form a saturated compound. The ethene molecule can be represented as preparing for reaction thus:

$$\begin{array}{ccc} \overset{\displaystyle H}{\underset{\displaystyle H}{>}}C=C\overset{\displaystyle H}{\underset{\displaystyle H}{<}} & \longrightarrow & H-\overset{|}{\underset{|}{C}}-\overset{|}{\underset{|}{C}}-H \\ & & \quad\; H\; H \end{array}$$

Here are eight different ways in which ethene can form addition compounds – nearly all of which are of industrial importance:

(1) *Plus hydrogen* (with heated nickel catalyst) ⟶ ethane, C_2H_6
(2) *Plus oxygen* (silver catalyst at *c.* 250 °C) ⟶ ethene oxide, C_2H_4O
(3) *Plus chlorine* ⟶ dichloroethane, $C_2H_4Cl_2$
(4) *Plus bromine* ⟶ dibromoethane, $C_2H_4Br_2$

(5) *Plus hydrogen chloride* (*c.* 200 °C) $\longrightarrow$ monochloroethane, C_2H_5Cl

(6) *Plus water* (steam at 300 °C, 60atm., orthophosphoric acid catalyst)
$\longrightarrow$ ethanol, C_2H_5OH

(7) *Plus other ethene molecules* $\longrightarrow$ polythene, $(C_2H_4)_n$

(8) *Plus benzene* (C_6H_6) $\longrightarrow$ ethyl benzene, $C_6H_5C_2H_5$

PROCESS (1): This is known as **hydrogenation.** Moderate temperatures and pressures are used and any alkene can be converted to the corresponding alkane in this way. Nickel is always the catalyst:

$$C_2H_4 + H_2 \longrightarrow \underset{\substack{| \\ H}}{\overset{\substack{H \\ |}}{H-C}}-\underset{\substack{| \\ H}}{\overset{\substack{H \\ |}}{C}}-H$$

PROCESS (2): Ethylene, when burnt, has the usual reaction of a hydrocarbon giving carbon dioxide and water:

$$C_2H_4 + 3O_2 \longrightarrow 2CO_2 + 2H_2O$$

For complete combustion, therefore, it needs three times its own volume of oxygen (fifteen times its own volume of air), and burns with a more luminous and smoky flame than methane.

At a lower temperature, with silver as catalyst, an oxide is produced with the structure:

$$H-\overset{\substack{O \\ /\!\!\backslash}}{\underset{\substack{| \\ H}}{C}}-\underset{\substack{| \\ H}}{C}-H$$

(Any organic compound that has in its structure an oxygen atom linking two carbon atoms in this way is known as an 'epoxy' compound. The term is in everyday use for the epoxy-resins used in paint manufacture, adhesives, etc.)

This oxide is a volatile liquid and can be converted into ethane-diol (commonly known as 'ethylene glycol'), $C_2H_4(OH)_2$, either by treatment with steam under pressure at 200 °C or with dilute hydrochloric acid at room temperature. This causes the addition of a molecule of water to give the well known 'anti-freeze' liquid:

$$\overset{\substack{O \\ /\!\!\backslash}}{H_2C-CH_2} + H_2O \longrightarrow \overset{\substack{OH\ OH \\ |\quad |}}{H_2C-CH_2}$$

PROCESSES (3) AND (4): Give simple addition compounds that are now regarded not as dihalides of ethene but as substitution compounds derived from ethane, e.g. dichloroethane rather than ethene dichloride:

$$\underset{\substack{| \\ H}}{\overset{\substack{| \\ H}}{H-C}}-\underset{\substack{| \\ H}}{\overset{\substack{| \\ H}}{C}}-H + Cl_2 \longrightarrow \underset{\substack{| \\ H}}{\overset{\substack{Cl \\ |}}{H-C}}-\underset{\substack{| \\ H}}{\overset{\substack{Cl \\ |}}{C}}-H$$

PROCESS (5) Hydrogen chloride is added (hydrochlorination) giving monochloroethane

$$
\begin{array}{c}
\;\;\;\; | \;\;\; | \\
\mathrm{H-C-C-H} + \mathrm{HCl} \longrightarrow \\
\;\;\;\; | \;\;\; | \\
\;\;\;\; \mathrm{H}\;\;\;\mathrm{H}
\end{array}
\qquad
\begin{array}{c}
\;\;\;\mathrm{H}\;\;\;\mathrm{Cl} \\
\;\;\;\; | \;\;\; | \\
\mathrm{H-C-C-H} \\
\;\;\;\; | \;\;\; | \\
\;\;\;\; \mathrm{H}\;\;\;\mathrm{H}
\end{array}
$$

PROCESS (6): This is the modern industrial method for making ethanol:

$$
\begin{array}{c}
\;\;\;\; | \;\;\; | \\
\mathrm{H-C-C-H} + \mathrm{H_2O} \longrightarrow \\
\;\;\;\; | \;\;\; | \\
\;\;\;\; \mathrm{H}\;\;\;\mathrm{H}
\end{array}
\qquad
\begin{array}{c}
\;\;\;\mathrm{H}\;\;\;\mathrm{OH} \\
\;\;\;\; | \;\;\; | \\
\mathrm{H-C-C-H} \\
\;\;\;\; | \;\;\; | \\
\;\;\;\; \mathrm{H}\;\;\;\mathrm{H}
\end{array}
$$

Fermentation methods produce an impure dilute aqueous solution of ethanol from which it is difficult to produce the pure alcohol (Section 59).

PROCESSES (7) AND (8): See Section 57.

The laboratory preparation of ethene (Section 75) is the reverse of process (6).

To show that ethene is an unsaturated compound two simple tests can be used:

(a) shaken with a little bromine water, the brown colour disappears as ethene is converted to dibromoethane;
(b) shaken with potassium manganate(VII) solution (acidified with a few drops of dilute sulphuric acid) the solution becomes colourless as the ethene is oxidized to colourless ethane-diol (ethylene glycol).

It also forms an ozonide with ozone by addition (Section 30). Ethene is an endothermic compound, its heat of formation being $+11.3\,\mathrm{kJ\,mol^{-1}}$.

Propene (propylene), C_3H_6

This has the structural formula

$$
\begin{array}{c}
\mathrm{H_3C} \qquad\quad \mathrm{H} \\
\diagdown \qquad \diagup \\
\mathrm{C=C} \\
\diagup \qquad \diagdown \\
\mathrm{H} \qquad\quad \mathrm{H}
\end{array}
$$

where one of the hydrogen atoms of ethene has been replaced by a methyl group, the double bond between the carbon atoms remaining. Propene has chemical properties similar to those of ethene and from it can be made a polymer 'polypropylene' which in some ways is more useful than 'polythene' (Section 57).

ALKYNES

The simplest and best-known alkyne is acetylene, now called ethyne. Its molecular formula is C_2H_2 with the structure $\mathrm{H-C{\equiv}C-H}$. The next compound in this homologous series is propyne, C_3H_4, with the structure $H_3C-C\equiv C-H$. Butyne, the next member, has two isomers $H_5C_2-C\equiv C-H$ and $H_3C-C\equiv C-CH_3$. From these examples it can be seen that the general formula for all members of this series is C_nH_{2n-2}.

Ethyne was formerly obtained almost entirely by the action of water on calcium carbide, and this is still the laboratory method (Section 75). Calcium carbide is manufactured by heating calcium oxide and coke at 2000 °C in an electric furnace – so its production is centred in areas where electric power is cheap, e.g. Norway and Niagara:

$$CaO + 3C \longrightarrow CaC_2 + CO$$

Calcium carbide is needed, not only for the production of ethyne but also for making calcium cyanamide ('nitrolim(e)') by its union with nitrogen in another electric furnace process. Nitrolime, CaNCN, is a valuable, water-soluble nitrogenous fertilizer that has a theoretical nitrogen content as high as that of ammonium nitrate (35% by mass):

$$CaC_2 + N_2 \longrightarrow CaNCN + C$$

The other product is graphite.

As carbon is tetravalent, CaC_2 seems a strange formula. Ethyne, although insoluble in water, has some acidic properties and can form salts by replacement of its hydrogen atoms by a metal. Divalent calcium replaces two hydrogen atoms giving

$$Ca^{2+}(C\equiv C)^{2-}$$

Recently ethyne has been obtained, as in N. Italy, by the breakdown of natural gas under special conditions. Then, with ethyne as the starting point, important compounds such as ethanal (acetaldehyde), ethanoic acid (Section 56), and PVC (Section 57) can be obtained.

Alkynes are even more unsaturated than the alkenes, a triple bond existing between the carbon atoms. Alkynes respond to the same tests for unsaturation given for ethene earlier in this section. Ethyne is extremely endothermic, its heat of formation being $+200\,kJ\,mol^{-1}$ and under some conditions it can be very dangerous. It cannot be stored in cylinders under pressure without danger of explosion unless it is dissolved in an inert solvent. It burns with a very luminous flame but, unless a specially-designed jet is used, it produces much carbon black. Ethyne explodes with chlorine gas.

Besides its importance as a source of other valuable organic compounds, ethyne is required for use in oxy-acetylene welding and steel cutting. The temperature produced by an oxy-acetylene blow-pipe can reach 3000 °C.

BENZENE

Benzene, C_6H_6, is the most important organic ring-structure carbon compound. (Compounds that have at least one benzene ring in their structure are known as **aromatic** compounds.) Benzene is a colourless liquid (b.p. 80 °C), a very good solvent for many covalent compounds, and is as flammable as petrol. As it is easily absorbed through the skin and may cause cancer, no risks should be taken with it.

Its empirical formula is known to be CH and its relative molecular mass, 78 – which gives the molecular formula C_6H_6. What then is its structural formula? If it were an unsaturated straight-chain hydrocarbon it would be capable of adding on another eight monovalent atoms thus:

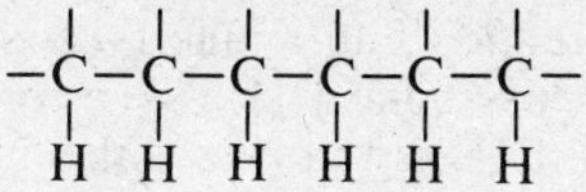

$$-\overset{|}{\underset{|}{C}}-\overset{|}{\underset{|}{C}}-\overset{|}{\underset{|}{C}}-\overset{|}{\underset{|}{C}}-\overset{|}{\underset{|}{C}}-\overset{|}{\underset{|}{C}}-$$
$$\quad H \quad H \quad H \quad H \quad H \quad H$$

No-one, however, could add on more than six and it was Kekulé, in 1865, who first suggested the ring structure – which is now known to be a regular hexagon in shape:

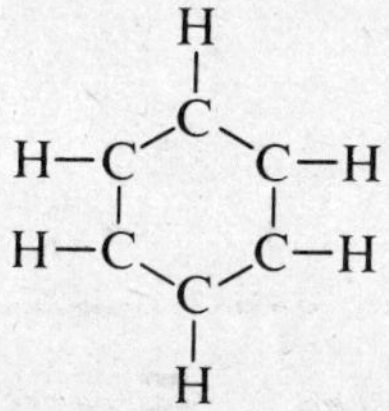

This formula still leaves each carbon atom with only three valence bonds and each carbon atom has one electron that is not being used. To represent this, Kekulé suggested that alternate double bonds should be shown as in Figure 31, but the more modern practice is to draw a circle inside the hexagon.

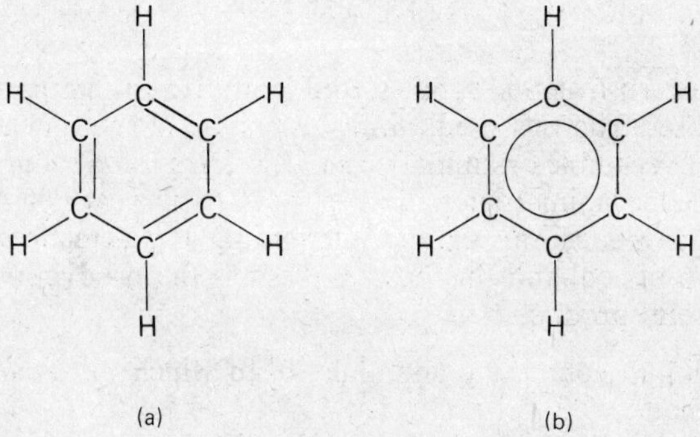

(a)   (b)

Figure 31

Benzene has many of the properties associated with unsaturation and it will form some addition compounds. For example:

Hydrogenation gives the compound C_6H_{12}, cyclohexane.
Chlorination gives $C_6H_6Cl_6$, benzene hexachloride (BHC) – a powerful insecticide sold under various trade names.
With ozone it forms a tri-ozonide (an ozone molecule adding at each double bond).

Benzene also forms important substitution compounds. If one of the hydrogen atoms is replaced by a methyl group, the next member of the homologous series is obtained, **toluene**, C_7H_8 ($C_6H_5.CH_3$). If two methyl groups substitute two hydrogen atoms, the **xylenes** are obtained ($C_6H_4.(CH_3)_2$, and three isomers are possible). If a hydroxyl group is introduced in place of a hydrogen atom, **phenol**, C_6H_5OH, is formed. This has some of the properties of an alcohol, but it is also a weak acid. Its everyday name is 'carbolic acid' and it exists as colourless crystals which are corrosive to the skin. In very weak aqueous solution it is used as an antiseptic. If an amino ($-NH_2$) group takes the place of one hydrogen atom, phenylamine, $C_6H_5NH_2$, commonly known as **aniline** is obtained. The radical $-C_6H_5$ is known as the 'phenyl' radical and this is sometimes indicated by 'Ph' (not to be confused with the pH of a solution which has a small p and a large H). Aniline is the starting point for a whole range of aniline dyes.

Benzene, and other common aromatic compounds, including phenol and aniline, used to be obtained exclusively from coal tar, but now they can be derived from petroleum.

55 Flame. Explosion. Fire Extinguishers

FLAME

In earlier times, rush lights, candles, and primitive oil-lamps were used for lighting purposes. The oils used, until the middle of the nineteenth century, were always of vegetable or animal origin, e.g. colza (oil obtained from seeds), sperm, and melted animal fat (tallow). In the early years of the nineteenth century, coal gas made its appearance for lighting. It was made by the destructive distillation of coal and this process, besides the production of gas, gave three other useful products:

(1) a watery liquid containing ammonia, from which nitrogenous fertilizers could be made;
(2) coal tar, from which hundreds of valuable organic compounds (mostly aromatic) could be made;
(3) a residue of coke.

In some parts of Europe, where there was no coal but plenty of wood, destructive distillation of wood gave a useful gas that burnt with a rather more luminous flame, the other products being charcoal, wood tar, and a watery liquid which, unlike that from coal, was acid in solution. From this watery liquid other useful organic compounds could be obtained: ethanoic acid, methanol, and acetone. The destructive distillation of wood is still done on a small scale for obtaining special types of charcoal and the by-products are not wasted.

Gas lighting was much improved when Welsbach, in 1885, invented the gas mantle and from then on lighting depended on the heating power of the gas and not on its illuminating power. The mantle consisted of a delicate fabric of cerium and thorium oxides that emitted bright light when it was strongly heated, i.e. it became **incandescent.** Coal gas was modified to make it more efficient and less poisonous giving different varieties of **town gas.** Town gas has, in turn, given way to natural gas. During the twentieth century, the use of gas for lighting has steadily diminished and it is now mainly used for fires, cookers, and central heating.

When one 'lights' a gas fire, to obtain the flame one has to use a spark or lighted match, because the chemical action between gas and air will not start until its **ignition temperature** is reached.

Flame is the boundary between gases (or vapours) where chemical action produces light and heat.

Flames, therefore, are hollow and there is little heat inside them. Furthermore, flame has a reciprocal nature because the same flame will be produced either if gas A is allowed to pass into gas B or if gas B is allowed to pass into gas A. As we live in an atmosphere of diluted oxygen, natural gas provides us with flame. If, however, we lived in an atmosphere of natural gas, oxygen would have to be supplied to obtain flame. Simple experiments illustrate (a) the hollowness of flame and (b) the reciprocal nature of flame:

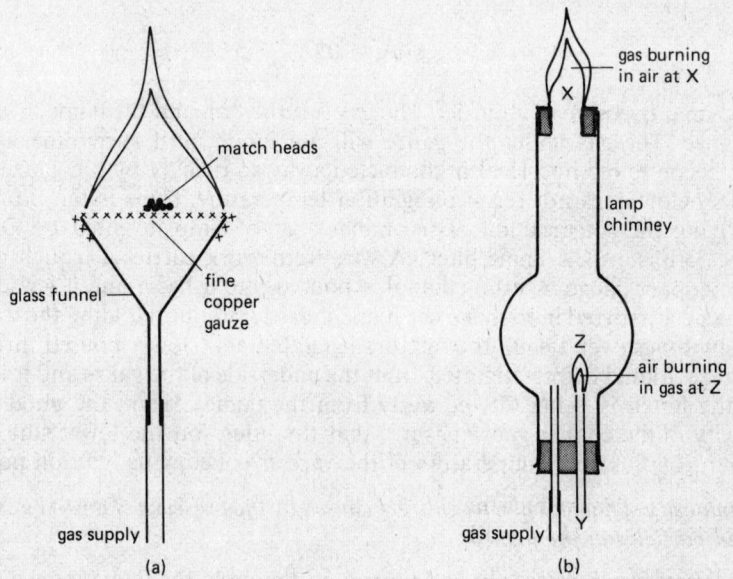

Figure 32

In Figure 32(a) a gauze of fine copper wire is stretched across a glass funnel and in the middle of the gauze are placed the heads of a few matches. Gas (town or natural) is passed into the funnel and the flame is lit at the top. The match heads do not ignite.

In Figure 32(b) a lamp chimney is set upright and, at its lower end, it is closed by a cork through which one tube admits gas and another tube, shorter and thinner, admits air. The top of the chimney is fitted with a cork bearing a short length of glass tubing about 3 cm in diameter. The tube at the top, X, is covered, the gas turned on, and when all the air has been swept out, the gas is lit at Y. The top is now uncovered and the gas ignited at X. Immediately the flame at Y recedes and burns at the top of the short tube, Z. Gas is now burning in air at X and air is burning in gas at Z. (A little damp sodium chloride smeared round the rim of the short tube at Z before the experiment will help the small flame there to be more clearly seen.)

Two simple experiments will show that a definite ignition temperature has to be reached before a flame will appear. Figure 33(a) shows an ordinary gauze

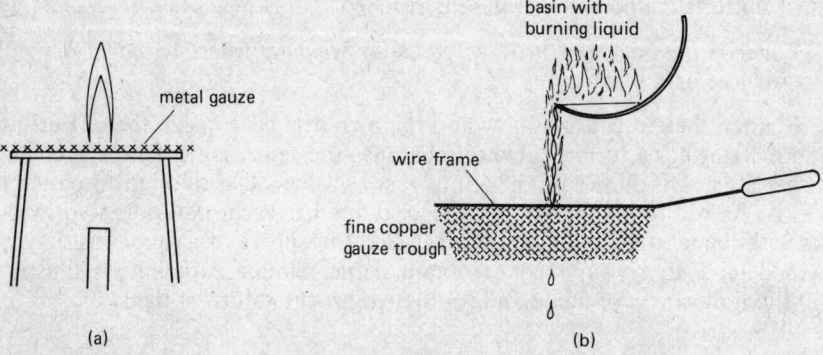

metal gauze

basin with
burning liquid

wire frame

fine copper
gauze trough

(a)   (b)

Figure 33

placed on a tripod over a burner. The gas is turned on and the flame lit *above* the gauze. The gas under the gauze will not ignite until such time as the gauze becomes red hot. Heat is conducted away so quickly by the gauze that the gas below does not reach its ignition temperature. (This is the principle underlying the construction of the miner's safety lamp invented by Davy.) Figure 33(b) shows a 'flame filter'. A wire framework carries a trough made of fine copper gauze. A little ethanol is poured into a basin and a few drops of toluene are stirred in to make the flame more luminous. Holding the framework by tongs over a sink, the mixture is ignited and slowly poured through the gauze. Liquid drips, unignited, from the underside of the gauze and it looks as if the liquid is being filtered away from the flame. Again, the good conductivity of the copper gauze ensures that the liquid on the lower side does not burn because the temperature of the vapour is below its ignition point.

Luminosity of flame is almost entirely caused by the presence of incandescent solid particles in the flame.

If a laboratory burner is lit and tapped on the desk, the flame immediately becomes more luminous because of the dislodging of dust, chalk particles, etc. It does not matter whether the particles are combustible or not. Again, if a basin of cold water is lowered into the flame of a burning hydrocarbon, the luminosity almost disappears and carbon black is seen on the underside of the basin. Temperature and pressure also affect luminosity to a slight extent. The flame of pure hydrogen burning at reduced pressure produces heat but practically no light.

EXPLOSION

A jet of hydrogen will burn quietly in a gas jar of oxygen or in a gas jar of chlorine, but if hydrogen is mixed with oxygen or chlorine before the mixture is ignited, a violent explosion occurs. The difference lies in the fact that if the gases are mixed before ignition, the chemical reaction is not confined to a boundary (as in flame) but can occur between the molecules over the whole volume simultaneously, with the rapid production of heat and consequent expansion of gas. Most explosions, other than atomic ones, are caused by the sudden production and expansion of gases. Gunpowder, for example, consists of a mixture of 75% potassium nitrate, 14% charcoal, 10% sulphur, and about 1% water. This mixture will burn quietly in the open air, but if detonated in a confined space it will explode. In the latter case, oxygen provided by the nitrate produces oxides of carbon and sulphur which rapidly expand with the heat of reaction. Endothermic compounds are usually explosive because they rapidly decompose with the evolution of heat.

A treacle tin experiment well illustrates the difference between flame and explosion. The lid and base of a treacle tin are each punctured in the centre with a hole of about 5 mm diameter. The lid is removed and the tin is filled with town or natural gas by the displacement of air. The lid is replaced, the tin placed on a tripod, and the gas lit at the top. The flame is luminous at first but then develops two cones (the inside one being unburnt gas) and becomes smaller. All the time, inside the tin, the gas burnt at the top is being replaced by air entering though the hole at the bottom and, eventually, when the tin is full of the right proportions of gas and air, an explosion wave travels into the tin and an explosion over the whole volume of the mixed gases results, blowing the lid high in the air.

The experiment also illustrates what happens when a laboratory burner 'strikes back'. If air is admitted at the bottom of the burner tube in order to obtain a hotter flame (as in the Bunsen and Meker type burners), the tube contains an explosive mixture of gas and air. If the pressure on the gas supply is high enough, the flame remains at the top of the tube. If, however, the gas tap is turned to lower the gas pressure when the air hole(s) is open wide, an explosion wave can travel down the tube and ignite the gas at the issuing jet. If this happens it should be corrected at once; otherwise the metal tube will become very hot. A sharp blow on the rubber tube connecting the burner, after the gas supply has been fully restored, will usually put things right by causing a slight interruption of the gas flow.

FIRE EXTINGUISHERS

The chief object of any type of fire extinguisher is to exclude air from the substance that is burning. There are five main types now in use, each of which has advantages in particular circumstances:

(1) ACID-BICARBONATE: One part of the apparatus contains solid sodium hydrogencarbonate, $NaHCO_3$. The other part has a vessel containing hydrochloric acid. When put into use, the acid container is broken and from the nozzle a fizzing mass, with much carbon dioxide is immediately forced out. ($NaHCO_3$ gives a higher percentage by mass of carbon dioxide (52.4%) than any other common carbonate or hydrogencarbonate.)

(2) FOAM: (See Section 42 under aluminium sulphate.)

(3) ORGANIC LIQUID: A liquid is chosen that is non-flammable and has a high relative molecular mass. Its heavy vapour 'blankets' the material and excludes air. Tetrachloromethane, CCl_4, has been used in the 'Pyrene' type extinguishers. Its vapour has a relative density (at s.t.p.) of 77, but the toxicity of its vapour is a drawback. Nowadays mixed halogen derivatives of methane are preferred because they are less toxic, more efficient, and have even denser vapours. 'BCF' would indicate a bromochlorofluoro- derivative, e.g. $CBrClF_2$. In this compound, one hydrogen atom of the methane molecule has been replaced by a bromine atom, one by a chlorine atom, and two by fluorine atoms; the relative density of its vapour (at s.t.p.) is about 82. Organic liquid extinguishers must *not* be used on sodium fires.

(4) POWDER: Particularly useful for small domestic fires. The fine white powder is almost certain to contain, among other things, sodium hydrogen-carbonate and borax. When heated, the former gives steam and carbon dioxide and the latter swells remarkably giving off steam that is held in a bubbling froth.

(5) SOLID CARBON DIOXIDE: Useful for dealing with laboratory fires.

---

56   Alcohols. Ethers. Organic Acids. Esters

ALCOHOLS

The molecule of any alcohol contains at least one hydroxyl group, $-OH$, the functional group. If this group replaces one hydrogen of a methyl group and this is linked to a radical, R, then a **primary alcohol** is obtained, $R.CH_2OH$.

If another hydrogen atom of the methyl group is replaced by a second radical (which can be the same as, or different from, the first) a **secondary alcohol**,

$$\begin{matrix} R \\ R \end{matrix}\!\!\bigg\rangle CHOH$$

results. If the third hydrogen is replaced by another radical, a **tertiary alcohol**,

$$\begin{matrix} R \\ R - C(OH) \\ R \end{matrix}$$

is obtained. As there can also be alcohols with more than one hydroxyl group in the molecule, there are large numbers of such compounds. In Section 54, 'ethane-diol' (ethylene glycol) was mentioned which is a **dihydric alcohol,**

$$
\begin{array}{c}
CH_2OH \\
| \\
CH_2OH
\end{array}
$$

Glycerol, mentioned later in this section is a **trihydric alcohol,**

$$
\begin{array}{c}
CH_2OH \\
| \\
CHOH \\
| \\
CH_2OH
\end{array}
$$

There are two common primary alcohols: methanol (methyl alcohol), $H.CH_2OH$ (or CH_3OH), and ethanol (ethyl alcohol), $H_3C.CH_2OH$ (or C_2H_5OH). Ethanol is the alcohol well known as a constituent of alcoholic drinks. Absolute (pure) ethanol is seldom seen. For domestic use 'methylated spirit' (about 90% ethanol with poisonous material and colouring matter) is available without excise duty. For laboratory work 'rectified spirit' (about 95% ethanol), which is colourless but non-potable (unfit for drinking) is available. For medical work 'surgical spirit' (95% ethanol with antiseptics) is used. Methanol is much more poisonous than ethanol. The introduction of one or more hydroxyl groups into an organic molecule tends to make the compound more soluble in water. Methanol and ethanol are both completely miscible with water but for the higher alcohols, where the radical accounts for the greater part of the relative molecular mass, they become increasingly less soluble.

Typical reactions of ethanol

SODIUM: A primary alcohol can be regarded as water with one of its hydrogen atoms replaced by an organic radical, $R-OH$. Sodium reacts with it much as it does with water, but the reaction is more gentle:

$$2C_2H_5OH + 2Na \longrightarrow 2C_2H_5ONa + H_2$$

Hydrogen gas and sodium ethoxide are formed.

CONCENTRATED SULPHURIC ACID: There can be three quite different reactions according to the conditions:

(1) acid in excess gives *ethene* when warmed;
(2) the two compounds in molar proportions gives an *ester*;
(3) ethanol in excess gives an *ether*.

CHLORIDES OF PHOSPHORUS (Section 66): Another reaction showing a parallel between ethanol and water. An organic halide is produced. The reaction between phosphorus(III) chloride and ethanol produces monochloroethane (ethyl chloride) and phosphorous acid (Section 66):

$$3C_2H_5OH + PCl_3 \longrightarrow 3C_2H_5Cl + H_3PO_3$$

OXIDATION : If alcohols are oxidized, they first lose hydrogen to form compounds called **aldehydes** and then gain oxygen to form **acids.** The word 'aldehyde' is derived from '*al*cohol *dehyd*rogenated'. The functional group of an aldehyde is

$$-C\diagup^{H}_{\diagdown\!\!\!_O}$$

where the oxygen and hydrogen atoms are separately linked with the carbon atom, unlike the hydroxyl group $-O-H$.

The oxidation of ethanol gives the following:

$$H_3C.CH_2OH \xrightarrow{-2H} H_3C.C\diagup^{H}_{\diagdown\!\!\!_O} \xrightarrow{+O} H_3C.C\diagup^{OH}_{\diagdown\!\!\!_O}$$

ethanol	ethanal	ethanoic acid
(ethyl alcohol)	(acetaldehyde)	(acetic acid)

According to the systematic naming of organic compounds, the names of all alcohols and phenolic compounds end in '-ol'; all aldehydic compounds end in '-al'; all carboxylic acids end in '-oic acid'. As the three compounds above are derivatives of ethane, the names are readily understood.

The oxidizing agent for the above stages is the dichromate ion, $Cr_2O_7^{2-}$. If ethanol is mixed with dilute sulphuric acid and either sodium or potassium dichromate solution the mixture, when heated in a distilling apparatus, gives ethanal vapour which can be condensed in a well-cooled receiver. The boiling point of ethanal is $21\,^{\circ}C$ – only just above normal room temperature.

Most aldehydes are acrid-smelling liquids which have many interesting reactions that are beyond the scope of this book. Methanal (formaldehyde), derived from methanol, is a gas (b.p. $-21\,^{\circ}C$). Its solution in water 'formalin' is well known in biological laboratories. Molecules of aldehydes readily group together to form large molecules – polymers – and 'meta' solid fuel is a polymer of ethanal. All aldehydes reduce Fehling's Solution (Sections 58 and 72).

ETHERS

An ether has a functional group of one oxygen atom joining the carbon atoms of two radicals, $R-O-R$. The radicals can be the same or different. If they differ, the ether is known as a 'mixed' ether, methoxyethane (methyl ethyl ether), $H_3C-O-C_2H_5$, being an example. The commonest ether, the vapour of which has been used as an anaesthetic, is ethoxyethane (diethyl ether), $H_5C_2-O-C_2H_5$ or simply $(C_2H_5)_2O$. It is prepared in the laboratory by carefully mixing ethanol and concentrated sulphuric acid in molar proportions (46 parts ethanol to 98 parts acid by mass). The mixture is placed in a distilling flask fitted with a tap funnel so that more ethanol can be continuously added whilst the mixture is warmed on a sand bath. Ethoxyethane distils as follows:

1st stage (molar proportions)

$$C_2H_5OH + H_2SO_4 \rightleftharpoons C_2H_5HSO_4 + H_2O$$

2nd stage (more ethanol added)

$$C_2H_5HSO_4 + C_2H_5OH \longrightarrow (C_2H_5)_2O + H_2SO_4$$

Ethanol and ethoxyethane are two of the most important solvents used in organic chemistry. Ethers are dangerously flammable but, apart from this, they are almost as chemically inert as the alkanes. Ethoxyethane is very volatile (b.p. 34.5 °C) and has been the cause of serious explosions with oxygen in operating theatres where the strictest precautions have not been observed. In the laboratory, 'ether' should not be used anywhere near naked flames, hot gauzes or electric filaments. It has a heavy vapour which can creep along a bench.

ORGANIC ACIDS

The functional group of a carboxylic acid is

This group, known as the **carboxyl** group, has one more oxygen atom than has the aldehydic group. Nearly all organic acids have molecules containing at least one carboxyl group, its representation usually being condensed to −COOH. Phenol (Section 54) and other phenolic compounds do, however, also behave as weak acids.

The first three members of the homologous series of acids derived from the alkanes are:

H.COOH	H₃C.COOH	H₅C₂.COOH
methanoic	ethanoic	propanoic
(formic)	(acetic)	(propionic)
acid	acid	acid

These are all **monocarboxylic** acids because the molecules contain only one acid group. Oxalic acid

$$\begin{array}{c} COOH \\ | \\ COOH \end{array}$$

is a **di**carboxylic acid.

Methanoic (formic) acid, H.COOH

This is one of the few really unpleasant organic acids. It blisters the skin and is found in the stings of ants and bees. (Its common name is derived from the Latin 'formicum' meaning 'ant'.) Its boiling point 100.5 °C is almost the same as that of water and when it is dehydrated it gives pure carbon monoxide (Section 75).

Ethanoic (*acetic*) *acid,* $CH_3.COOH$

This acid has been known for centuries (in dilute aqueous solution) as vinegar, which is obtained whenever wine goes sour (from the French 'vin aigre' meaning 'sour wine') (Section 59). It would not, however, be worth while recovering the acid from this impure dilute solution. In the laboratory it can be obtained by prolonged oxidation of ethanol with acid dichromate solution. Industrially, there are now two important ways of producing it (1) from ethene and (2) from ethyne:

(1) Ethene and steam combine at 300 °C with the help of high pressure and a catalyst (Section 54) to give ethanol:

$$C_2H_4 + H_2O \longrightarrow C_2H_5OH$$

(2) Ethyne, now readily available from natural gas, combines with water in the presence of mercury(II) sulphate as catalyst and 30% sulphuric acid, to form ethanal:

$$C_2H_2 + H_2O \longrightarrow CH_3CHO$$

Ethanal, or ethanol, can then be oxidized, by air under pressure of about 5 atm. with a manganese(II) salt as catalyst, to give the acid.

Large quantities of ethanoic acid are needed industrially for the manufacture of cellulose ethanoate, photographic film, aspirin, various esters (see below), metal ethanoates, as a precipitating agent when obtaining latex from rubber or casein from milk, and as a solvent.

Ethanoic acid has a boiling point of 118.2 °C and the vapour burns with a blue flame, rather like that from ethanol. Its melting point is 16.7 °C and in cold weather the pure acid freezes solid as colourless crystals – 'glacial acetic acid'. Concentrated ethanoic acid is only slightly ionized and is a poor conductor. On dilution, its conductance steadily improves as the extent of ionization increases.

ESTERS

Whenever an alcohol and an acid combine an **ester** and water are formed. The acid can be either organic or inorganic (ethyl hydrogensulphate, $C_2H_5HSO_4$ produced from molar proportions of ethanol and concentrated sulphuric acid, is an ester). Esters are sweet-smelling liquids some of which are used for essences and flavourings, e.g. amyl ethanoate (pear-drop flavour); ethyl butanoate (pineapple); ethyl methanoate (rum flavour).

To obtain an ester, the acid and alcohol are heated together in the presence of concentrated sulphuric acid, or some other powerful dehydrating agent. Without the dehydrating agent the action is reversed because esters are easily hydrolysed.

Esterification

Alcohol + Acid $\rightleftarrows$ Ester + Water

Hydrolysis

It used to be thought that the function of the concentrated sulphuric acid was simply to remove the water formed and hence prevent the 'back' reaction. It is now known that the reaction is catalysed by hydrogen ions; so the acid acts also as a catalyst. Hydrogen chloride can bring about the same result:

$$C_2H_5OH + CH_3.COOH \rightleftharpoons CH_3.COOC_2H_5 + H_2O$$

The full structural formula of ethyl ethanoate is:

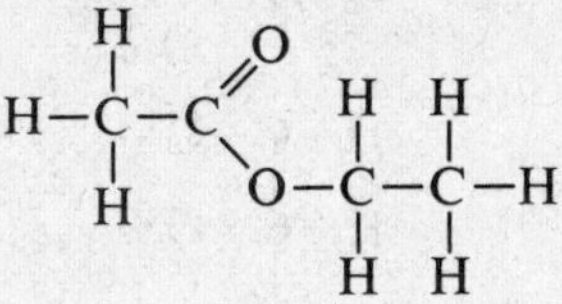

Isomerism (Section 53) is common among esters. The ester ethyl ethanoate is isomeric with another ester derived from propanoic acid – methyl propanoate, $C_2H_5.COOCH_3$. The two radicals involved have simply exchanged places.

Ethyl ethanoate is used as a solvent for paints and varnishes and amyl ethanoate as a solvent for nitrocellulose.

The most convenient way of hydrolysing an ester is to heat it with sodium hydroxide solution. The action of water is too slow and the reaction would not go to completion. When the alkali hydrolyses ethyl ethanoate the products are ethanol and sodium ethanoate, not the acid itself. The alcohol can be distilled from the sodium salt in solution:

$$CH_3.COOC_2H_5 + NaOH \longrightarrow CH_3.COONa + C_2H_5OH$$

If methyl propanoate were treated in the same way, methanol would distil. This gives a simple way of distinguishing between the two isomers.

Soap making

This involves the hydrolysis of esters. The hydrolysis, in this case, is known as **saponification** (from the Latin 'sapo' meaning 'soap' and 'facio' meaning 'I make'). Most animal and vegetable oils or fats are mainly composed of esters derived from a well-known trihydric alcohol (i.e. the molecule contains three hydroxyl groups) called glycerol (glycerine). The structural formula of glycerol is:

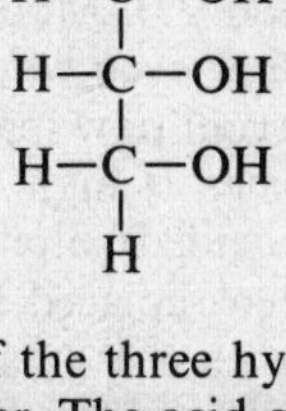

The hydrogen atom of each of the three hydroxyl groups can be replaced by an acid group to form an ester. The acid groups are mainly derived from three monocarboxylic acids (one carboxyl group per molecule) of high relative molecular mass: palmitic acid, $C_{15}H_{31}.COOH$; stearic acid, $C_{17}H_{35}.COOH$; and oleic acid, $C_{17}H_{33}.COOH$. The general formula of the saturated monocarboxylic acid series (starting with methanoic) is $C_nH_{2n+1}.COOH$ so that it can be seen that both palmitic and stearic acids are members of this series. Oleic acid, having two hydrogen atoms less in its molecule than stearic acid, is an unsaturated acid with a double bond between two of the carbon atoms. If the stearic group $C_{17}H_{35}CO-$ is represented for simplicity as *St*, the tristearate of glycerol will have the formula:

$$
\begin{array}{c}
H \\
| \\
H-C-OSt \\
| \\
H-C-OSt \\
| \\
H-C-OSt \\
| \\
H
\end{array}
$$

If this compound is hydrolysed with sodium hydroxide solution, the two products will be sodium stearate (three moles) and glycerol (one mole). Soap is composed of the sodium salts of acids such as the three mentioned and therefore the saponification of fats yields two important products – soap and glycerol. When the fat and alkali have been boiled together, brine is added and the soap separates as solid curds. This is then purified, perfumed, and made into tablets. If soap is dissolved in ethanol and then recovered from this, the solid is almost transparent. If the original fat had been boiled with potassium (instead of sodium) hydroxide, 'soft' soap would have been obtained.

A good detergent (cleansing agent) has three properties: wetting power, dirt-removing power, and also the power to emulsify and suspend the particles of dirt. Detergent molecules have high relative molecular mass and at one end of each molecule there is a hydrophobic ('water-fearing') group and at the other end a hydrophilic ('water loving') group such as $-OH$ or $-COO^-$. The hydrophobic end is attracted to the surface of the material to be cleaned and the hydrophilic part attracts water molecules to ensure that the material is properly wetted. *All detergents are surface-acting agents – commonly called 'surfactants'.* The hydrophobic part of the molecule may be positive and the hydrophilic end negative, but the whole molecule is neutral. Fats and grease are soluble in the hydrophobic part of the molecules and are hence removed. The charges on the detergent help finally to suspend the loosened dirt particles so that water can remove them.

Soap consists of long-chain molecules of which the $C_{15}H_{31}-$ or $C_{17}H_{35}-$ groups constitute the hydrophobic part and the $-COONa$ the hydrophilic part. The disadvantage of soap is that it is rendered ineffective by the ions of calcium and magnesium (Section 27). Modern synthetic detergents are not affected and will lather with water even when these ions are present.

Two important types of detergent now made are the alkyl aryl sulphonates derived from benzene and alkenes of high relative molecular mass and the alkyl sulphates. The latter are called sulphates because one hydrogen of a sulphuric acid molecule has been replaced by a sodium atom and one by a long carbon chain. 'Alkyl' implies the presence of an aliphatic radical and 'aryl' the presence of an aromatic one (see Section 54 under 'Benzene').

57   Plastics and Polymers

PLASTICS

A plastic is a substance that, under the right conditions, can be moulded into any required shape.

A **thermosetting plastic** is one that, having been heated and moulded, will harden on cooling and, thereafter, no amount of heating will soften it again. A **thermoplastic** compound will soften when heated and lose its shape but will become rigid again when cool; a cycle that can be repeated indefinitely. All plastics are polymeric substances.

Landmarks in the development of plastics

1855:   First synthetic plastic made by the action of nitric acid on cellulose (carbohydrate (Section 58) made from cotton) – *cellulose nitrate* sold under trade names such as 'xylonite' and 'celluloid'. It is thermoplastic and a comb made of celluloid will soften and lose its shape when placed in hot water. It is also dangerously flammable and in the early days of the cinema serious accidents resulted from the use of celluloid film. Ping-pong balls made of it burn furiously.

1894:   Non-flammable *cellulose acetate* was made. Now used for wrapping material, 'non-flam' film, recording tape, and lacquers.

1897:   Plastics derived from *casein* (the chief protein substance in cow's milk). Still used for buttons, buckles, and, in fibre form, for blending with cotton, wool, etc.

EARLY TWENTIETH CENTURY: 'Bakelite' made from methanal (formaldehyde) and phenol. Thermosetting and brown in colour. Good insulator and was much used for electric-light fittings.

1926:   *Urea-formaldehyde* plastics that were colourless and could be dyed in bright colours. Used for beakers, buckets, dishes, bathroom fittings.

1930s: A flood of organic polymers such as polythene, polystyrene, polyvinyl chloride (PVC), polymethylmethacrylate (perspex), and the polyamide – nylon.

1940 ONWARDS: Polypropylene, polyurethanes, and terylene have been added.

POLYMERS

A polymer is formed when two or more molecules of one compound join together to make a larger unit, nothing being lost in the process.

The polymer has properties that are different from those of the original simple compound (monomer). Two molecules together form a 'dimer', three molecules a 'trimer', and so on. Most important plastic materials have hundreds of molecules joined together and the relative molecular masses of such compounds run into thousands.

The simplest unsaturated compound to polymerize is ethene. Here the double bond between the carbon atoms is disrupted and the unused valence electron of each carbon atom can be used to unite other similar molecules in a very long chain:

$$H_2C:CH_2 \longrightarrow \begin{array}{cc} H & H \\ | & | \\ -C-C- \\ | & | \\ H & H \end{array}$$

monomer

$$\begin{array}{cccccccccc} H & H & H & H & H & H & H & H & H & H \\ | & | & | & | & | & | & | & | & | & | \\ -C-C-C-C-C-C-C-C-C-C- \\ | & | & | & | & | & | & | & | & | & | \\ H & H & H & H & H & H & H & H & H & H \end{array} \quad (C_2H_4)_n$$

polymer

Having once started, the polymerization process will continue until the addition of, say, methyl groups, $-CH_3$, stops the ends. Polythene (polyethylene) was discovered by accident. A chemist was attempting to make ethene combine with another compound by using very high pressure. He did not succeed but found, when the apparatus was finally opened, a white waxy solid that was new to him. Further research showed that under high pressure, and with the second compound acting as a catalyst, ethene had polymerized into a white solid. Later the polymer was produced much more cheaply when the German chemist, Ziegler, discovered that certain complex organo-metallic compounds would catalyse the change at relatively low pressures. Polythene is now used for the manufacture of many cheap domestic articles. It is unaffected by most acids and alkalis, so reagent bottles, stoppers, etc., are often made of it. Being thermoplastic, however, bowls made from it will soften in very hot water.

Even if one of the hydrogen atoms of the ethene molecule is replaced by another atom or group, it may still be possible to polymerize it to give a polymer with different properties. Thus PVC, polypropylene, polystyrene, and polyacrylonitrile can be obtained.

Polyvinyl chloride (PVC)

Vinyl chloride is made by the direct combination of ethyne and hydrogen chloride with the help of a catalyst:

$$HC \equiv CH + HCl \longrightarrow \begin{array}{cc} H & Cl \\ | & | \\ -C-C- \\ | & | \\ H & H \end{array} \longrightarrow (H_2C \cdot CHCl)_n$$

vinyl chloride monomer   polymer

Vinyl chloride polymerizes to form polyvinyl chloride (PVC). This polymer has largely replaced rubber and lead in domestic electric wiring. It is thermoplastic and is also used for the manufacture of light raincoats.

Polypropene (polypropylene)

Propene is produced in large quantities by the cracking of alkanes of high relative molecular mass. It has the ethylene structure with one hydrogen replaced by a methyl group:

$$-\underset{\underset{H}{|}}{\overset{\overset{H}{|}}{C}}-\underset{\underset{H}{|}}{\overset{\overset{CH_3}{|}}{C}}-$$

propene monomer

Polypropene is rather more expensive than polythene. It softens at a higher temperature than polythene does and will withstand boiling water. It is also harder and less easily scratched than polythene.

Polystyrene

Styrene is made in two stages. First, ethene is combined with benzene (with aluminium chloride as catalyst) to form ethyl benzene, $C_6H_5.C_2H_5$. Second, the ethyl benzene is dehydrogenated with the help of heat and another catalyst, to give styrene:

$$C_6H_5.C_2H_5 - 2H \longrightarrow -\underset{\underset{H}{|}}{\overset{\overset{H}{|}}{C}}-\underset{\underset{H}{|}}{\overset{\overset{C_6H_5}{|}}{C}}-$$

styrene monomer

Polystyrene is thermoplastic and transparent. It is used for making unbreakable 'glass', insulators, ceiling tiles, floor covering, and artificial leather.

Polyacrylonitrile

Here one hydrogen atom of the ethylene molecule is replaced by a cyanide group. The compound is made by the addition of hydrogen cyanide, HCN, to ethyne:

$$HC\equiv CH + HCN \longrightarrow -\underset{\underset{H}{|}}{\overset{\overset{H}{|}}{C}}-\underset{\underset{H}{|}}{\overset{\overset{CN}{|}}{C}}-$$

acrylonitrile monomer

Polyacrylonitrile is used in making the fibre 'Orlon'.

Nylon and Terylene

Nylon and **terylene** are not strictly polymers, because in the reactions that produce them simple molecules are eliminated: HCl in the case of nylon and H_2O in the case of terylene.

NYLON: This was developed in 1938 by the Dupont de Nemours firm in the U.S.A. It was the first useful synthetic fibre, having been obtained by Carothers some years before. Until then all artificial fibres had been produced by the treatment of natural cellulose. Nylon contains nitrogen and its structure resembles that of a protein material (Section 60). It is a 'polyamide' and part of the chain shown below indicates its complicated structure:

$$-HN(CH_2)_6NHCO(CH_2)_4CONH(CH_2)_6NH-$$

TERYLENE: This was first made by British chemists in 1941 and has been developed by Imperial Chemical Industries Ltd. It is manufactured by the union of *tere*phthalic acid and ethane-diol (ethy*lene* glycol) (Section 56), the italicised syllables showing how the name was derived. Phthalic acid is an aromatic dicarboxylic acid, $C_6H_4(COOH)_2$. (The indicator, phenolphthalein, is made from this acid and phenol.) There are, however, three possible isomers with this formula and terephthalic acid is the one where the two carboxyl groups are diametrically opposite to one another in the benzene ring.

Both nylon and terylene give very strong hard-wearing fibres. Nylon rope is very much tougher than rope made from hemp or flax. Synthetic fibres generally, have properties that are different from those of natural fibres such as cotton, silk, and wool. Some of the differences are advantageous, some are not. It is usually the practice to blend synthetic with natural fibres in order to obtain the best of both worlds.

58   Carbohydrates: Sugars; Starch; Cellulose

These compounds are not well named because they are in no sense hydrates of carbon. They cannot be made by direct union of carbon with water. All carbohydrates, however, have the general formula $C_x(H_2O)_y$ so that the number of hydrogen atoms in the molecule is always twice the number of oxygen atoms. Sometimes x and y are so large (as in starch and cellulose) that they have not been finally determined and these are represented as $(C_6H_{10}O_5)_n$ where n is a very large number. Not every compound whose formula can be expressed in the form given above is necessarily a carbohydrate. Ethanoic acid $CH_3.COOH$ could be written as $C_2(H_2O)_2$ but it is an acid and not a carbohydrate.

SUGARS

In modern chemical nomenclature, the names of all sugars end in '-ose'. The familiar cane- (or beet-) sugar is sucrose, $C_{12}H_{22}O_{11}$. This can be converted quite easily into two other sugars – glucose and fructose – which each have the formula $C_6H_{12}O_6$. These are sometimes known as hexoses because they have six carbon atoms and are isomeric with one another. Sugars have many hydroxyl groups in their molecules and, unlike most organic compounds, are quite soluble in water.

Sucrose

This is extracted from crushed sugar-cane or sugar-beet. The soluble material is extracted with water and slaked-lime (calcium hydroxide) is added to neutralize any acid present and to precipitate protein material. Carbon dioxide is then passed through the solution to precipitate excess calcium ions as insoluble carbonate. No acid must be left because this would tend to hydrolyse the sucrose into hexoses. The liquid is finally filtered, decolourized with charcoal, and crystallized. The mother-liquor that does not crystallize is known as 'molasses'. Cane sugar comes mainly from the West Indies. Much sugar-beet is grown in England – chiefly in East Anglia. If the West Indian sugar is not decolourized it is sold as brown sugar – such as Demerara – but it should be remembered that white sugar crystals can be made to have any colour with the help of various colouring agents.

If crystals of sucrose are heated gently, there is some dehydration giving a thick pleasant-smelling dark-brown liquid known as 'caramel'. Strong heating, or reaction with concentrated sulphuric acid, dehydrates it completely leaving carbon ('sugar-char'). Sucrose solution warmed with Fehling's solution gives no reduction. No copper(I) oxide forms (Section 72).

Sucrose is used as a food, for preserving fruit, for jam making, and for the preparation of hexoses. Molasses can be used to make treacle, rum, silage for cattle food, or it can be converted to industrial ethanol.

Glucose and fructose

If sucrose is hydrolysed it is changed into a mixture of equimolar quantities of two simple sugars which are isomeric – glucose and fructose:

$$C_{12}H_{22}O_{11} + H_2O \longrightarrow C_6H_{12}O_6 + C_6H_{12}O_6$$
$$\text{sucrose} \qquad\qquad\qquad \text{glucose} \qquad \text{fructose}$$

The hydrolysis is brought about, both on a laboratory and an industrial scale, by warming sucrose solution with a very dilute acid (hydrochloric or sulphuric). Of the two sugars, glucose is the more important. Fructose (from the Latin 'fructus' meaning 'a fruit') occurs naturally in many fruit juices and is sometimes known as 'fruit sugar'.

In the molecular structure of glucose there are known to be five hydroxyl groups and one aldehydic group, so glucose, unlike sucrose, has some of the properties of an aldehyde. It will reduce Fehling's solution, when the two are heated together, to copper(I) oxide.

When the more complicated carbohydrates, such as sucrose and starch, are present in food that is eaten, they are broken down with the help of enzymes (organic catalysts) present in the digestive juices, to glucose. Glucose, then, with the help of insulin produced by the pancreas, is absorbed into the blood stream giving a prime source of energy. For this reason, glucose is used in large quantities in sweet making and an athlete can take it about a half-hour before special energy demands are made on his system. A person whose pancreas is not functioning as it should may suffer from diabetes. The hexose is not properly assimilated and appears in the urine, a condition easily confirmed by the use of Fehling's solution. A diabetic can lead a normal life if he takes insulin – usually by injection – that has been obtained from an ox. The balance between the insulin taken and the sugar content of the diet has to be carefully controlled.

Enzymes are catalysts that are produced by living organisms. Many enzymes act as hydrolytic ferments. It is now known that the organism need not necessarily be alive and the enzyme is a complicated compound that can be extracted from the organism. Enzyme names always end in '-ase' (zymase, invertase, etc.) or '-in' (pepsin, rennin, etc.). In the human saliva, starch is broken down with the help of two such agents – diastase and ptyalin.

In the U.S.A. enzymes are used on a large scale to produce glucose from maize producing the well-known 'corn syrup'.

STARCH $(C_6H_{10}O_5)n$

This occurs naturally in large quantities in foods such as potatoes, rice, grains of all kinds, and in every green plant. It is made by photosynthesis (Section 46) and can be stored in seeds and tubers. The formula and structure varies with the source but the molecule is known to consist of long chains (both straight and branched) of glucose units. When obtained from natural sources starch is in the form of granules that are insoluble in water, but a colloidal dispersion can be prepared (Section 28) which, when cold, gives the characteristic blue colour with an iodine solution (Section 23).

Like sucrose, starch can be hydrolysed by heating with dilute acid, but the process takes longer. Starch is changed first to an isomer of sucrose called maltose (malt sugar) and then to glucose:

$$(C_6H_{10}O_5)_n \xrightarrow{\;H_2O\;} C_{12}H_{22}O_{11} \xrightarrow{\;H_2O\;} C_6H_{12}O_6$$

Starch does not reduce Fehling's solution.

CELLULOSE $(C_6H_{10}O_5)n$

In this case the value of n is estimated to be between one hundred thousand and five hundred thousand. Cellulose exists in nature in even greater quantity than does starch, as it is the chief constituent of the cell walls of plants. Fibres of cotton, jute, hemp, and flax can contain up to 90% cellulose. Various woods contain up to 50%. Like starch, cellulose can be hydrolysed to glucose with the help of dilute acid. The enzyme that brings about the same change – cellulase – is not present in human digestive juices, so cellulose cannot be used for food by man. Only wood-eating animals, such as the 'silver fish', possess this enzyme in their digestive systems.

Cellulose is insoluble in water and also in organic solvents such as ethanol and ethoxyethane. It is an extremely important raw material for the manufacture of products such as paper, textiles, film, and explosives.

Paper: Wood is cut into small pieces and boiled with a solution of either sodium hydroxide or calcium hydrogensulphite, $Ca(HSO_3)_2$. This removes the non-cellulose material and produces a pulp from which various grades of paper can be made. Good-quality filter paper is almost pure cellulose.

Textiles: These include the spinning of natural fibres giving cotton material (from the cotton plant), linen (from flax), and the artificial silks known as 'rayon'. There have been various methods for turning cellulose into artificial silk, but the difficulty has been the discovery of a suitable solvent for cellulose so that threads can be drawn. One method uses sodium hydroxide solution with carbon disulphide, which produces a compound called cellulose xanthate ('viscose rayon'); in another method cellulose ethanoate (acetate) is dissolved in acetone ('acetate rayon').

Film: 'Non-flam' film is made from cellulose ethanoate.

Plastics: Cellulose ethanoate and cellulose nitrate (Section 57).

Explosives: Cellulose nitrate is obtained by treating cellulose with 'mixed acid', a mixture of nitric and sulphuric acids – both concentrated. If the reaction is carried to completion 'gun cotton' is produced. Gun cotton with glyceryl nitrate (nitroglycerine) gives the explosive 'cordite' and with glyceryl nitrate, wood pulp, and potassium nitrate gives 'gelignite'.

59   Fermentation: Beer; Wine; Spirits; Vinegar

If yeast is placed into a solution of sugar, or a dispersion of starch, at room temperature it will feed on the carbohydrate with the evolution of carbon dioxide, which causes the slow effervescence so characteristic of **fermentation.** Yeast is a simple unicellular organism which, when it has developed to a certain size, buds and produces more cells which break away from the parent organism. At about 35 °C the process is quite rapid and if further nourishment is added to provide phosphorus and nitrogen (e.g. a little ammonium orthophosphate) the development is even more remarkable. Associated with yeast there always exist various enzymes that catalyse the breakdown processes. Starch is first converted to maltose, then to glucose, and finally to ethanol and carbon dioxide. Brewer's yeast is a culture of the yeast plant and is kept under refrigeration until needed. Cold does not kill yeast but simply causes it to remain dormant. Baker's yeast is much less active because it contains yeast cells compressed in a damp starchy medium.

BEER

This is made by fermentation of 'malt liquor'. Moist barley grains are allowed to sprout in a clean atmosphere and during the germination an enzyme – diastase – is produced. The grains are then roasted to stop the germination, the product being known as 'malt'. This is made into a mash with water ('wort') and kept at a temperature of about 60 °C. The diastase converts most of the starch to the sugar maltose, $C_{12}H_{22}O_{11}$. The liquid is separated from insoluble material and boiled with hops to give it the characteristic bitter flavour. Finally the yeast is added at a temperature of about 20 °C. The fermentation takes place in large barrel-shaped containers so that the carbon dioxide formed can be taken off and used for gasifying bottled beer. During the fermentation, the enzyme maltase converts the maltose into glucose:

$$C_{12}H_{22}O_{11} + H_2O \xrightarrow{\text{maltase}} 2C_6H_{12}O_6$$
$$\text{maltose} \qquad\qquad\qquad \text{glucose}$$

Another enzyme, zymase, then breaks down the hexose into ethanol and carbon dioxide:

$$C_6H_{12}O_6 \xrightarrow{\text{zymase}} 2C_2H_5OH + 2CO_2$$

The final liquor contains up to 7% ethanol, depending upon the time that fermentation is allowed to proceed. The beer is finally filtered and then either bottled with carbon dioxide gas or stored in barrels which are now commonly made of aluminium.

Ethanol is toxic to nearly all living organisms and no fermentation process can produce a liquor containing more than 15% ethanol, because the yeast cells cannot live in a greater concentration. It is not easy to determine the proportion of alcohol in a given liquid by means of an ordinary hydrometer. Ethanol has a relative density of about 0.8 and it might be assumed that if equal volumes of ethanol and water were mixed together the resultant relative density would be 0.9. This is not the case because when these two liquids are mixed there is always a shrinkage of volume. Special hydrometers accompanied by tables have to be used when assessing the alcoholic content of any aqueous solution.

Fermentation is *not* used as a method of obtaining pure ethanol because the process would be too complicated and costly.

WINE

Wine is made by the fermentation of grape juice but, in this case, no yeast is added because the ferments required are present in the 'bloom' on the grape itself. If grapes are left long on the vine, fermentation will begin of its own accord. The longer the grapes are left, the more sugar is produced and sweet white wines are made from grapes that are picked very late in the season. Normally, when the grapes are ripe enough, they are picked, crushed and allowed to ferment at the right temperature. If red grapes are lightly crushed and the skins removed immediately, white wine can be produced. If some colour is allowed from the grape skins, the pink (rosé) wine will result. Crushed and left in contact with the skins for a period, the grapes will produce red wine. White grapes naturally produce a white wine. Grape sugar is glucose, so the fermentation proceeds as shown in the equation above. The ethanol content of ordinary wine varies from 5–12%.

SPIRITS

To produce alcoholic liquors of high ethanol content, the fermented liquor must be distilled. This eventually produces a product of about 60% ethanol which will obtain special flavour either from its source, from added flavouring or from both of these. Irish whiskey, derived from the starch of potatoes, has a different flavour from Scotch whisky obtained from malt liquor. In the Far East, spirits are made by the fermentation of starch from rice, followed by distillation. Distillation of fermented grape juice gives brandy. Gin is usually a grain spirit flavoured with the juniper berry. Vodka is also a grain spirit that has very little aroma or flavouring. When spirits are matured by keeping for a number of years in casks, the flavour and 'bouquet' improve through the formation of small quantities of various esters, but the alcoholic content slowly drops. Ethanol has a boiling point of 78 °C compared with 100 °C for water; so ethanol evaporates more quickly than water.

Fortified wines

If a bottle of wine is opened and left exposed to the air, organisms from the air soon begin to oxidize the alcohol into ethanoic acid and the wine goes sour. If the ethanol content is more than 15% this cannot happen; so extra ethanol is added, often in the form of brandy, to raise the alcohol content to 20–25%. Such 'fortified' wines will keep for some time without deterioration after opening. They include sherry, madeira, port and vermouth.

VINEGAR

This is obtained by the deliberate souring of a dilute ethanolic liquor with the aid of a fungus called *Mycoderma aceti*. In England the vinegar sold is usually malt vinegar, because the stages of its preparation follow the same lines as those for the brewing of beer. The fermentation is carried further than for beer and then the ethanol-containing liquid is trickled over beech shavings, held in huge casks, on which the *Mycoderma aceti* is growing. Beech is chosen because it does not give the unpleasant flavour to the product that a wood like oak would. Plenty of air has to be admitted to the cask whilst this further fermentation proceeds; otherwise some ethanal (acetaldehyde) (Section 56) would be formed giving the vinegar an unacceptable taste.

Any liquid containing 10–15% ethanol can be used to make a vinegar. In

France wine vinegar is common and in Scotland many prefer vinegar prepared from diluted whisky. In Devonshire it can be made from cider, which itself has been made by the fermentation of apple juice. A passable imitation of vinegar can be made by adding caramel (to give the brown colour) to a dilute solution of ethanoic acid, but legally this must not be called 'vinegar'. Vinegar contains a rather higher percentage of ethanoic acid than the original liquid contained of ethanol. A mole of ethanol (46 g) should give a mole of ethanoic acid (60 g).

60   Proteins

Proteins are organic nitrogenous compounds of very high relative molecular mass. They nearly always contain some sulphur, but those that contain phosphorus in their composition are known as phosphoproteins. A large part of all living matter consists of proteins but most of them are difficult, or so far impossible, to synthesize. Proteins make up the fibrous, insoluble matter of hair, wool, natural silk, skin, etc. Soluble proteins include egg albumen and gelatin, though they form colloidal dispersions rather than true solutions (Section 28). Gelatin is used for the preparation of edible jellies and also for the coating of photographic film. Proteins play a very important part in the human diet in such foods as lean meat, eggs, and cheese. All enzymes are proteins.

Although proteins are difficult to synthesize, it is a comparatively easy matter to break them down into simpler compounds and the end product always consists of one or more amino acids. The simplest amino acid has the structural formula:

$$
\begin{array}{c}
H \\
| \\
H_2N-C-COOH \\
| \\
H
\end{array}
$$

This is known as aminoethanoic (aminoacetic) acid, also called glycine. It is an extremely interesting compound and most of its special properties derive from its functional groups. This acid can be looked upon as methane with one hydrogen of its molecule replaced by a basic (amino-) group, $-NH_2$, and one by an acid (carboxyl) group, $-COOH$. Consequently, the amino acids always have a dual nature – part base, part acid. Proteins themselves have this dual nature. Protein colloidal dispersions, for example, can be positively charged under some conditions but negatively charged under others. The first protein to have its structure elucidated was insulin – the substance that plays such an important part in the assimilation of sugar by the human body (Section 58).

It is now possible to synthesize protein material for use as animal feeding-stuff starting from simple compounds such as methanol (derived from North Sea gas) and ammonia.

61 Silicon (Group IV)

Silicon is next below carbon in Group IV of the Periodic Classification and there are many points of resemblance between the two. Unlike carbon, the element is seldom found in everyday life but in combination with other elements it is the second most common in the Earth's crust – second only to oxygen. All its compounds are covalent and, in them, silicon always has a valency of four. The commonest are the oxide (silica), SiO_2, of which ordinary sand is mainly composed, and various silicates such as clay. Sand and clay are the chief constituents of most soils. The element can be prepared in the laboratory by the reduction of its oxide with magnesium powder – a good example of the reducing power of an electropositive metal:

$$SiO_2 + 2Mg \longrightarrow 2MgO + Si$$

This is a very exothermic reaction. (Magnesium also burns in carbon dioxide giving magnesium oxide and carbon.) The magnesium oxide is removed with acid and the silicon remains.

Silicon is a hard, grey, crystalline solid which burns in oxygen at about 400 °C. It can combine at high temperatures with some other non-metals and its compound with carbon – silicon carbide, SiC – is almost as hard as diamond. It is not attacked by any acids except hydrofluoric acid, HF. It is, however, readily attacked by strong alkalis when warmed. Sodium hydroxide solution gives sodium silicate and hydrogen:

$$Si + 2NaOH + H_2O \longrightarrow Na_2SiO_3 + 2H_2$$

Formulae of the simple silicon compounds are analogous with those of the corresponding carbon ones, e.g. the hydrides:

alkanes CH_4, C_2H_6, etc.   :   silanes SiH_4, Si_2H_6, etc.

Calcium silicate occurs in the slag from the blast furnace in the extraction of iron (Section 71). Silicon itself is used to make some special steels. The element is a semi-conductor and, in very pure form, is used for transistors.

Atoms of silicon can, like those of carbon, catenate to form chains but after a linkage of about six atoms the chain breaks. There is not, therefore, an unlimited number of possible compounds as there is with carbon. A more stable linkage is that provided by alternate atoms of oxygen and silicon thus:

$$
\begin{array}{ccccccccc}
 & R & & R & & R & & R & & R \\
 & | & & | & & | & & | & & | \\
-O- & Si & -O- & Si & -O- & Si & -O- & Si & -O- & Si- \\
 & | & & | & & | & & | & & | \\
 & R & & R & & R & & R & & R
\end{array}
$$

This is the basic framework of all the valuable **silicones.** They can be obtained in liquid or solid form and most of them have various organic radicals (represented by R above) on the silicon atom's other valence bonds. They are all chemically inert and various silicones are manufactured for uses such as synthetic rubber, lubricants, greases, heating-bath fluids, water repellants for waterproofing fabrics, anti-foam preparations, etc. The greases and rubbers are not affected by very low temperatures.

Silicon oxide (silica), SiO_2

This is the only oxide of silicon. Although quite insoluble in water it *is* an acid oxide because it can be dissolved in a fused or concentrated solution of an alkali to give a salt (silicate) and water only:

$$SiO_2 + 2NaOH \longrightarrow Na_2SiO_3 + 2H_2O$$

A thick syrupy solution of sodium silicate is made by heating the solid with boiling water under pressure for some time. Known as 'water glass' it can be used for treating floor surfaces and as an egg preservative.

Because silica (sand) is an acid oxide it can unite with many basic oxides to form the various silicates found in nature. Clay is a complicated mixture of many compounds, but pure china clay is mainly aluminium silicate, $Al_2O_3, 2SiO_2, 2H_2O$.

If dilute hydrochloric acid is added to a sodium silicate solution, a white gelatinous precipitate of hydrated silica is obtained. If this is purified and carefully dehydrated **silica gel** is obtained in solid form. This is a very porous substance and is much used as a water absorbent for keeping foodstuffs dry. It is often also used in desiccators.

Pure silica is colourless, though ordinary sand is often coloured by traces of iron, etc. It melts at a high temperature and pieces of apparatus such as tubing, crucibles, and basins can be made from it; also containers used to hold reagents for photochemical reactions. It is not attacked by common acids and will withstand considerable 'thermal shock'. It has a very low coefficient of expansion and cold water can be poured onto red-hot silica without cracking the silica. It is therefore useful in the laboratory for the pyrolysis of nitric acid (Section 64). Silica ware is much more expensive than glass.

Glass

The origins of glass are lost in antiquity, but it is essentially a mixture of metallic silicates and has no definite melting point; it only softens over a fairly wide range of temperature. It is classified as a 'supercooled liquid' because it has no crystalline structure and when molten glass is cooled there seems to be insufficient time for it to take up a crystalline form. It is always translucent or transparent.

Common cheap 'soda-glass' has three main ingredients in its manufacture: soda, lime, and sand. Thus its main chemical constituents are sodium silicate and calcium silicate. Crude glass ('bottle glass') is coloured either green (caused by traces of iron(II)) or brown (traces of iron(III)). To decolourize it two things are needed: an oxidizing agent to make sure that all the iron is iron(III) and then something that will provide a pink tint to counteract the brown colour due to iron(III). For this purpose, manganese(IV) oxide has been used because it is an oxidizing agent and gives the necessary pink colour. Modern mixtures for removing colour contain sodium nitrate, arsenic(III) oxide, and a little selenium (to give the pink tinge).

In making glass, if potash is used instead of soda a product results that softens at a higher temperature and this is known as 'hard glass'. It is more costly and can only be manipulated if softened with a blow-pipe flame. For laboratory ware, glass tubing, expendable test-tubes, etc. soda-glass is used but ignition tubes of various sizes will be made of potash-glass or borosilicate-glass (see below).

Both soda- and potash-glasses are light to handle but do not have a very attractive appearance. They are not clear or bright, nor highly reflecting or refracting. In 1675, George Ravenscroft, trying to rival the imported Venetian glassware, discovered how to make the 'English glass of lead'. Instead of lime he used lead oxide (either litharge or red-lead) and obtained a denser and more brilliant product. In modern times, this type of glass is used for chandeliers, tableware, cut glass of all sorts, and optical glass.

An important recent development has been the manufacture of glass ovenware which withstands a much higher 'thermal shock' than other types of glass. Though not as good in this respect as silica itself, it will survive sudden changes of temperature without cracking. In its composition the percentage of silica is increased and oxides of boron and aluminium are added. The product is often known as 'borosilicate' glass.

The approximate percentage by mass of compounds present in the four main types of glass described above are as follows:

Soft (soda): SiO_2, 73; CaO, 11; Na_2O, 14; other compounds, 2.
Hard (potash): Much the same as above but K_2O instead of Na_2O.
Lead: SiO_2, 55; PbO, 32; K_2O, 13.
Borosilicate: SiO_2, 80; B_2O_3, 13; Na_2O, 5; Al_2O_3, 2.

Most laboratory glassware is now made of borosilicate glass. It will withstand much greater mechanical, as well as thermal, shock than can the cheaper soda-glass.

Glass can be coloured in various ways by adding to the melt up to 1% of various metallic oxides, for example:

Blue or green (according to amount added): copper or chromium
Blue: cobalt
Ruby-red: selenium
Amethyst: nickel
Orange: selenium with cadmium sulphide
White: tin(IV) oxide or barium sulphate

Vast quantities of green, red, and orange glasses are made for use in traffic lights.

Glass, like silica itself, is not attacked by acids except hydrofluoric acid. Alkalis attack glass very readily and strong alkalis must not be heated in glass vessels. Soda-lime (Section 75 under methane) is a useful substitute.

NITROGEN

Nitrogen is the major constituent of the air and its laboratory extraction from the air and its isolation from liquefied air have already been briefly described (Section 30). The preparation of pure nitrogen from a nitrite and an ammonium salt is dealt with in Section 75.

The nitrogen molecule, N_2, is particularly stable and this makes the element very unreactive. It combines with some electropositive metals, such as magnesium, to form nitrides and it will react directly with hydrogen to form ammonia and with oxygen to form nitrogen monoxide, NO. The last two reactions take place only under extreme conditions but they are very important in 'fixing' nitrogen, i.e. in making it into useful compounds. There is no specific test for nitrogen; its presence must be proved by the elimination of other possibilities. The major use of nitrogen in industry is the manufacture of ammonia from which other nitrogen compounds of importance can be made (Section 63).

OXIDES OF NITROGEN

Nitrogen forms six oxides but only three of these are common: dinitrogen oxide, nitrogen monoxide, and nitrogen dioxide. In none of these does nitrogen show its normal valence of 3.

Dinitrogen oxide, N_2O

Formerly known as nitrous oxide and as 'laughing gas', it is prepared by the pyrolysis of ammonium nitrate (Section 75). The gas readily decomposes into its elements when heated:

$$2N_2O \longrightarrow 2N_2 + O_2$$

For this reason it is a very good supporter of combustion and will even rekindle a glowing splint. The gaseous products of the reaction above contain one-third by volume of oxygen whereas air contains only one-fifth. Dinitrogen oxide can be distinguished from oxygen by testing with nitrogen monoxide. Only in the case of oxygen are brown fumes of the dioxide formed.

The gas is still used as an anaesthetic for some minor operations.

Nitrogen monoxide, NO

Formerly called nitric oxide, this gas is now sometimes known simply as nitrogen oxide. It is usually prepared in the laboratory by the reaction of nitric acid with copper turnings (Section 75). On exposure to oxygen or the air, it is immediately oxidized to the brown fumes of nitrogen dioxide:

$$2NO + O_2 \longrightarrow 2NO_2$$

Because of this reaction and the fact that nitrogen dioxide has a very pungent smell, it is impossible to find out whether nitrogen monoxide has a smell of its own.

Nitrogen monoxide will not support combustion of weakly burning materials such as sulphur or wood because it does not decompose much below 1000 °C. Fiercely burning substances such as phosphorus or magnesium, however, will decompose the gas and so continue to burn.

The gas is formed from its elements by passing a high-energy electrical discharge through air (Section 64):

$$N_2 + O_2 \longrightarrow 2NO$$

This occurs during thunderstorms and, since oxygen and water are also present in abundance, the nitrogen monoxide is converted to nitric acid and taken by rain into the soil.

Nitrogen dioxide, NO_2, and, at low temperatures, *dinitrogen tetroxide*, N_2O_4

Prepared by the action of heat on lead nitrate (Section 75). When liquid (b.p. 22 °C) the molecules are almost entirely N_2O_4 (pale yellow in colour) but, on warming, a yellow/brown gas containing N_2O_4 molecules is formed. On further heating, the N_2O_4 molecules progressively dissociate and the mixture steadily darkens in colour:

$$\underset{\text{yellow/brown}}{N_2O_4} \quad \rightleftharpoons \quad \underset{\text{dark brown}}{2NO_2}$$

At 140 °C the gas reaches its darkest colour and consists entirely of NO_2 molecules. The gas met in the laboratory under normal conditions is really an equilibrium mixture of dinitrogen tetroxide and nitrogen dioxide. Often, when 'nitrogen dioxide' is spoken of, it is this mixture that is meant. If the gas is heated above 140 °C it becomes paler again as there is further dissociation into nitrogen monoxide and oxygen. This is completed at about 620 °C when the mixture of gases has no colour (Section 48):

$$\underset{\text{dark brown}}{2NO_2} \rightleftharpoons \underset{\text{colourless}}{2NO} + \underset{\text{colourless}}{O_2}$$

Nitrogen dioxide supports combustion well, as the heat of burning substances decomposes it into a mixture containing one-third oxygen by volume – as may be deduced from the above equation. Thus the gas will rekindle a brightly glowing splint.

Nitrogen dioxide has a very pungent odour and should not be inhaled as it can cause septic pneumonia. It is readily absorbed by cold water forming nitric and nitrous acids:

$$2NO_2 + H_2O \longrightarrow HNO_3 + HNO_2$$

For this reason, it is called a mixed acid anhydride.

---

63   Ammonia and Ammonium Salts

AMMONIA

Ammonia is prepared in the laboratory by warming an ammonium salt with an alkali (Section 75). As ammonium salts are all derived from ammonia itself, a large-scale version of this method cannot be used.

Industrially, the gas is made from nitrogen and hydrogen by the *Haber* process:

$$N_2 + 3H_2 \rightleftharpoons 2NH_3 \qquad \Delta H = -92kJ$$

The nitrogen comes from the air and the hydrogen by the cracking of hydrocarbons. Unfortunately, under normal conditions, the reaction proceeds very slowly and the equilibrium position lies well over to the left. To hasten the reaction it is necessary to increase both pressure and temperature for reasons discussed in Section 47. A catalyst would also be helpful. Again, for a good equilibrium yield of ammonia it is necessary to use a high pressure but a low temperature. Thus a compromise (optimum) temperature must be chosen to ensure that the maximum quantity of ammonia is produced in a given time. In practice, a mixture of nitrogen and hydrogen in the proportions 1:3 by volume is passed over a catalyst of finely divided iron at 350–450 °C. The pressure chosen can be from 200 to 1000 atm. High pressures make the process more dangerous. The ammonia formed is separated from unreacted nitrogen and hydrogen either by liquefying it or by dissolving it in water.

Ammonia has a powerful, pungent smell. It causes watering of the eyes and, although it can be a stimulant in very low concentration in the air, in higher concentrations it is poisonous. It is more soluble in water than any other common gas. A saturated solution of ammonia in water, at room temperature, has a density of only $0.88\,g\,cm^{-3}$. (It is often referred to as '0.88 ammonia solution'.) The very high solubility of the gas in water is caused mainly by hydrogen bonding, but some formation of ions occurs:

$$NH_3 + H_2O \rightleftharpoons NH_4^+ + OH^-$$

For this reason, ammonia solution behaves as a weak alkali and reacts with acids to form ammonium salts.

Ammonia's great solubility in water can be demonstrated by the 'fountain' experiment. A dry flask fitted with a bung, which carries a tube ending in a jet, is filled with dry ammonia. The flask is inverted and the lower end of the tube is immersed in water coloured with red litmus solution. A little ethoxyethane is poured carefully over the flask to cool it by the evaporation of the ethoxyethane. Thus the contraction in volume of the gas coaxes a little water up the tube into the flask. As $1 \, cm^3$ of water is sufficient, under normal conditions, to dissolve about $1300 \, cm^3$ of ammonia, air pressure causes the flask to fill, the litmus turning blue at the same time. (Such an experiment can also be performed with other very soluble gases such as hydrogen chloride and sulphur dioxide, but in these cases blue litmus solution would be used initially.)

Many insoluble hydroxides are precipitated from solutions of salts by ammonia solution, familiar examples being:

$$Fe^{2+} + 2OH^- \longrightarrow Fe(OH)_2 \qquad \text{(dull-green precipitate)}$$
$$Fe^{3+} + 3OH^- \longrightarrow Fe(OH)_3 \qquad \text{(red-brown precipitate)}$$
$$Cu^{2+} + 2OH^- \longrightarrow Cu(OH)_2 \qquad \text{(pale-blue precipitate)}$$

In some cases salts and metal hydroxides react with ammonia to form **ammines**, in which ammonia molecules are attached to the metal ion by co-ordinate linkages (Sections 10 and 72). Anhydrous calcium chloride cannot be used for drying ammonia because it absorbs the gas forming the ammine $[Ca(NH_3)_6]^{2+}$.

Ammonia does not quite succeed in burning in ordinary air – it just gives a flicker of yellowish flame. If the air is enriched with oxygen, ammonia burns to form nitrogen and steam:

$$4NH_3 + 3O_2 \longrightarrow 2N_2 + 6H_2O$$

Catalytic oxidation: If ammonia and oxygen are mixed in the presence of a hot platinum catalyst, the ammonia is oxidized at a lower temperature producing not nitrogen but nitrogen monoxide:

$$4NH_3 + 5O_2 \longrightarrow 4NO + 6H_2O$$

This reaction is of great importance in the manufacture of nitric acid (Section 64). If a little 0.88 ammonia solution is placed in a small beaker, the space above will contain air mixed with ammonia gas. A hot platinum-wire spiral held a little above the liquid will become red-hot and continue to glow as the above exothermic reaction takes place. White fumes are seen – probably consisting of a smoke of ammonium nitrate formed by the reaction of excess ammonia with nitric acid derived from the nitrogen monoxide.

Ammonia is a good reducing agent, particularly for oxides of metals fairly low in the activity series. Its reduction of gently heated copper(II) oxide is a good example:

$$3CuO + 2NH_3 \longrightarrow 3Cu + N_2 + 3H_2O$$

Tests for the gas: Smell; turns red litmus blue; forms dense white fumes of ammonium chloride with hydrogen chloride gas.

Tests for the solution: Smell; turns red litmus blue; gives light-blue precipitate followed by deep-blue solution of ammine with copper(II) sulphate solution (Section 72).

Ammonia is used for the manufacture of nitric acid and the preparation of ammonium salts. It liquefies fairly easily (b.p. $-33\,°C$) and the liquid is used as a refrigerant in large cold stores and ice-cream factories – although it is no longer used in domestic refrigerators.

Liquid ammonia can now be injected directly into the soil as a high-nitrogen content fertilizer. It is fairly easily transported and it can be used as a convenient source of hydrogen because when 'cracked' thermally it gives a mixture of hydrogen and nitrogen in the ratio of $3:1$ by volume. For some of the uses of hydrogen the presence of the inert gas nitrogen may be an advantage.

$$2NH_3 \longrightarrow 3H_2 + N_2$$

Ammonia solution is used in the home as a water softener and cleansing agent.

AMMONIUM SALTS

Ammonium salts may be prepared by the reaction between ammonia and the appropriate acid in dilute solution, followed by evaporation and crystallization. They do not contain any water of hydration. Ammonium chloride may be prepared by the action of ammonia gas on hydrogen chloride gas. It appears as a white smoke that settles to a powder.

All ammonium salts yield ammonia when warmed with an alkali. Some of them, particularly those that are derived from volatile acids, dissociate on heating (Section 19). When the anion of the ammonium salt is derived from an oxidizing acid, oxidation products of ammonia are formed on heating and such actions are always irreversible. For example, ammonium dichromate gives chromium(III) oxide, nitrogen, and steam:

$$(NH_4)_2Cr_2O_7 \longrightarrow Cr_2O_3 + N_2 + 4H_2O$$

Similarly, the nitrate gives dinitrogen oxide and steam.

$$NH_4NO_3 \longrightarrow N_2O + 2H_2O$$

Ammonium sulphate, $(NH_4)_2SO_4$, is useful as a cheap nitrogenous fertilizer.

Ammonium nitrate, NH_4NO_3, is used as a fertilizer; also for explosives such as 'ammonal', and the preparation of dinitrogen oxide. As there is some danger of its exploding during storage, ammonium nitrate to be used as a fertilizer is often mixed with chalk, which renders it much safer. The mixture is known as 'nitro-chalk'.

Ammonium chloride, NH_4Cl, 'sal-ammoniac' is used in the manufacture of dry batteries.

64 Nitric Acid and Nitrates

NITRIC ACID

Nitric acid is best prepared in the laboratory by mixing a nitrate (usually potassium) with concentrated sulphuric acid and heating to drive off the volatile nitric acid (b.p. 86 °C):

$$KNO_3 + H_2SO_4 \longrightarrow KHSO_4 + HNO_3$$

As nitric acid corrodes cork or rubber an all-glass apparatus is used. The traditional apparatus is a retort with cooled receiver, as shown in Figure 34.

The flask fills with brown fumes, because some nitrogen dioxide is formed by the slight decomposition of the vapour and the distillate is coloured yellow

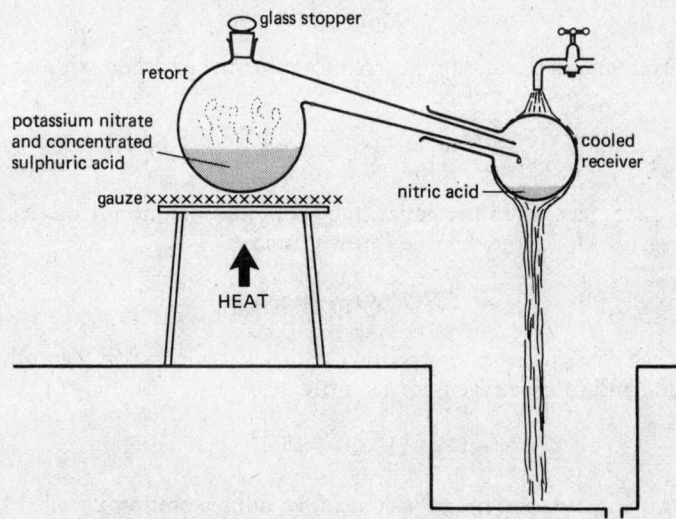

Figure 34 Preparation of nitric acid

by this gas dissolved in it. The dioxide can be removed by blowing air through the liquid. The almost 100% acid fumes in air and is known as 'fuming nitric acid'. The 'concentrated acid' of the laboratory is only about 69% acid, the rest being water.

Industrially, nitric acid is made by the Ostwald process which, for economic reasons, has ousted other methods. A mixture of ammonia (from the Haber process) and excess air, as a source of oxygen, is passed through a platinum/rhodium gauze which is initially heated by electricity to red-heat (Figure 35). After that, the exothermic reaction keeps the gauze at a temperature of about

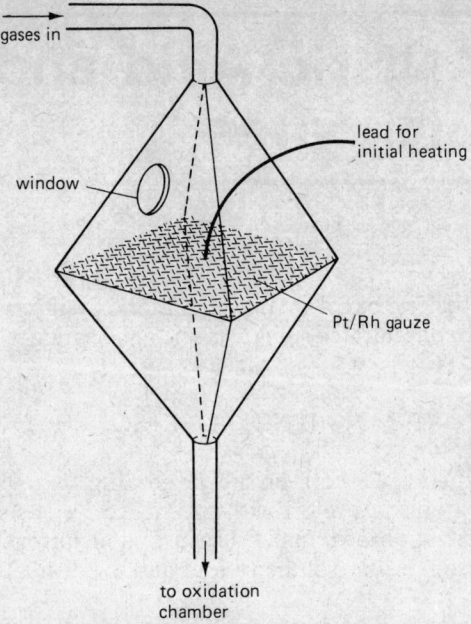

gases in

lead for
initial heating

window

Pt/Rh gauze

to oxidation
chamber

Figure 35

900 °C. Over 90% of the ammonia is oxidized to nitrogen monoxide and water, the gauze acting as a catalyst:

$$4NH_3 + 5O_2 \longrightarrow 4NO + 6H_2O$$

As the gases pass from the gauze, they cool and the monoxide reacts with more oxygen to form the dioxide (brown fumes):

$$2NO + O_2 \longrightarrow 2NO_2$$

The mixed gases pass up towers down which water is trickling and this, with the dioxide and more oxygen, forms nitric acid:

$$4NO_2 + O_2 + 2H_2O \longrightarrow 4HNO_3$$

Nitric acid is a strong monobasic acid. In dilute solution it will react with bases such as copper(II) oxide:

$$CuO + 2HNO_3 \longrightarrow Cu(NO_3)_2 + H_2O$$

It will also liberate carbon dioxide from carbonates. Nitrates are formed in all cases.

Concentrated nitric acid is also a strong oxidizing agent and, with the exception of gold and platinum, all common metals react with it. The reactions are complex and no equation can do more than indicate the chief products. Hydrogen is *not* formed, except in the case of very dilute acid with magnesium, where its oxidizing power is minimal. With the concentrated acid reactions, consideration of oxidation numbers can be very helpful.

Nitrogen has nine possible oxidation states and examples of all of them are known. The compounds in parentheses in Table 18 have been included to complete the table but they are not included in an elementary course of chemistry.

Table 18

Oxidation state		Compounds
+5	HNO_3	nitric acid
+4	NO_2	nitrogen dioxide
+3	HNO_2	nitrous acid
+2	NO	nitrogen monoxide
+1	N_2O	dinitrogen oxide
0	N_2	nitrogen
−1	$(H_2NOH$	hydroxylamine)
−2	$(N_2H_4$	hydrazine)
−3	NH_3	ammonia

Metals are oxidized by the acid and the acid itself is reduced. How far it will be reduced depends on several factors such as the concentration of the acid, the temperature, and how electropositive the metal is. A metal low in the activity series, such as copper, will reduce the acid to nitrogen dioxide or monoxide but an electropositive metal such as magnesium or zinc can produce dinitrogen oxide or even ammonia. If ammonia is formed locally in the presence of excess nitric acid, ammonium nitrate will result.

Because of its power as an oxidizing agent, fuming nitric acid can be dangerous. It will explode with some reducing agents such as white phosphorus or hot turpentine and even warm sawdust will burn violently.

The concentrated acid will oxidize sulphur (when warmed with it), hydrogen sulphide or sulphur dioxide to sulphuric acid:

$$S + 6HNO_3 \longrightarrow H_2SO_4 + 6NO_2 + 2H_2O$$
$$H_2S + 8HNO_3 \longrightarrow H_2SO_4 + 8NO_2 + 4H_2O$$
$$SO_2 + 2HNO_3 \longrightarrow H_2SO_4 + 2NO_2$$

With less concentrated acid, hydrogen sulphide is oxidized mainly to sulphur (white precipitate) and nitrogen monoxide is formed:

$$3H_2S + 2HNO_3 \longrightarrow 3S + 2NO + 4H_2O$$

Passed through a red-hot silica tube (or clay pipe) the concentrated acid readily pyrolyses into nitrogen dioxide, oxygen, and water. Using the apparatus shown in Figure 36, brown fumes are seen collecting in the boiling tube and the colourless gas collected over water will relight a glowing splint:

$$4HNO_3 \longrightarrow 4NO_2 + O_2 + 2H_2O$$

If a few drops of the concentrated acid are added to pale-green iron(II) sulphate, $FeSO_4$, solution, a dark-brown colour appears either immediately or upon warming. On heating further, effervescence occurs and the dark-brown

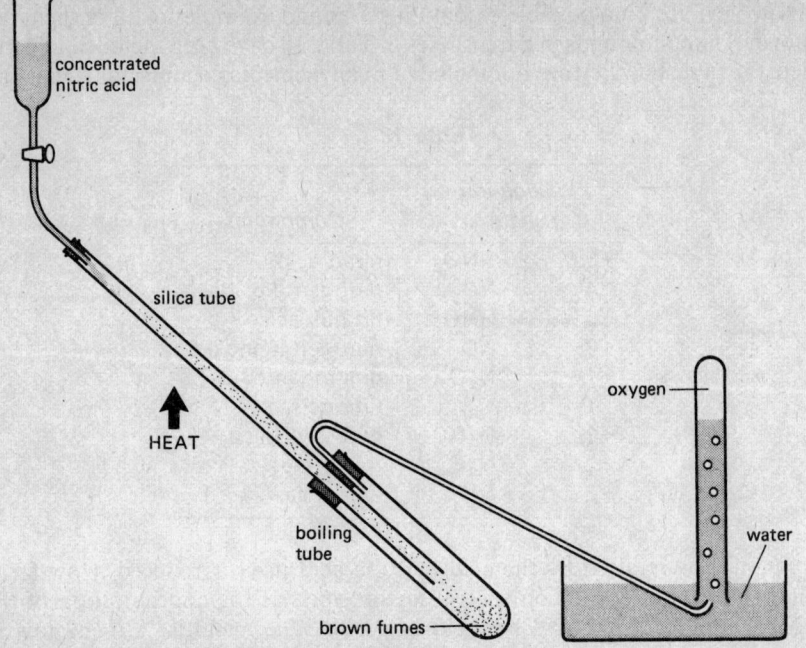

Figure 36   Pyrolysis of nitric acid

colour disappears leaving a pale-yellow solution. *Explanation:* First the nitric acid oxidizes iron(II) to iron(III) ions, itself being reduced to nitrogen monoxide:

$$4H^+ + 3Fe^{2+} + NO_3^- \longrightarrow 3Fe^{3+} + 2H_2O + NO$$

Nitrogen monoxide forms, by co-ordinate bonding with iron(II) ions that have not been oxidized, the nitroso-iron(II) ions, $Fe(NO)^{2+}$, which give the dark-brown colour. On heating, these split up into iron(II) ions and nitrogen monoxide once more. The dark-brown colour seen in the *brown-ring test* for nitrates is due to the $Fe(NO)^{2+}$ ions (Section 77).

Concentrated nitric acid corrodes the skin producing yellow stains, the colour of which is intensified if ammonia solution is added. The acid decomposes slowly under the influence of light, becoming yellow in colour, and is best stored in dark bottles.

The chief uses of nitric acid are in the manufacture of fertilizers (particularly ammonium nitrate), explosives, dyes, and plastics.

NITRATES

As all nitrates are soluble in water they cannot be prepared by precipitation methods but by the action of nitric acid upon the appropriate metal, oxide or carbonate (Section 37).

With all nitrates, heat causes pyrolysis. Considering the activity series given in Section 6, nitrates of the elements down to and including strontium decompose into the nitrite and oxygen, e.g. potassium nitrate:

$$2KNO_3 \longrightarrow 2KNO_2 + O_2$$

From calcium down to and including copper, decomposition of the nitrate is more complete giving the oxide, nitrogen dioxide, and oxygen, e.g. lead(II) nitrate:

$$2Pb(NO_3)_2 \longrightarrow 2PbO + 4NO_2 + O_2$$

Mercury and silver nitrates, when strongly heated (because their oxides are unstable) give the metal, nitrogen dioxide, and oxygen, e.g. mercury(II) nitrate:

$$Hg(NO_3)_2 \longrightarrow Hg + 2NO_2 + O_2$$

Ammonium nitrate, in a class of its own, gives dinitrogen oxide and steam:

$$NH_4NO_3 \longrightarrow N_2O + 2H_2O$$

If heated too rapidly, this salt may decompose with explosive violence. Brown fumes appear but the reactions involved are not understood.

Nitrates may be recognized by testing with concentrated sulphuric acid and copper (warmed) or by the brown-ring test (Section 77).

The main uses of nitrates are for fertilizers and explosives. Ammonium nitrate is an important nitrogenous fertilizer and vast deposits of sodium nitrate (Chile saltpetre) occur in Chile. Pure ammonium nitrate has 35% nitrogen by mass in its composition whereas sodium nitrate has only about 16.5%. Gunpowder contains potassium nitrate and ammonium nitrate is used in explosives such as amatol and ammonal.

65   The Nitrogen Cycle

All living things need nitrogen in order to build proteins (Section 60). Green plants use nitrates for this purpose which they can absorb in solution through their roots. Animals obtain their nitrogen compounds by eating plants or by eating other animals that have eaten plants. This process needs, therefore a continuous supply of nitrates from the soil. In nature, animals deposit their excreta, which contains nitrogen compounds, on the ground. Further, when plants and animals die they decay – a process caused by bacteria from the soil. Thus the nitrogen compounds, from excreta and dead organisms, are broken down into simpler ones and, eventually, into ammonium salts. Some nitrogen is lost (mainly as ammonia) to the atmosphere at this stage, as can be detected by the smell from some kinds of rotting manure. Other bacteria, the nitrifying ones, convert ammonium salts into nitrites and then, by further oxidation, into nitrates. Nitrogen loss is compensated, to a large extent, by two processes. First, thunderstorms produce some nitric acid which is washed into the soil (Section 62) and second, there are special bacteria which can live in nodules on the roots of leguminous plants (e.g. pea, bean, clover, vetch, lupin) which

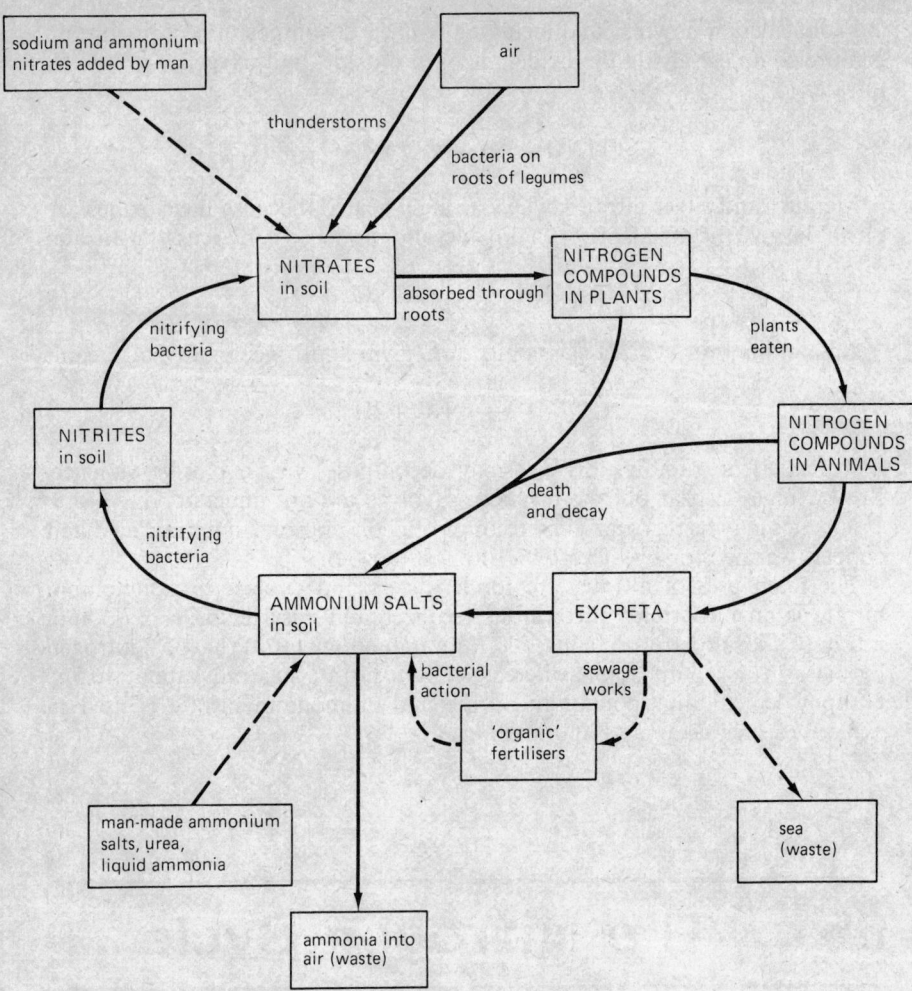

Figure 37   The nitrogen cycle

convert nitrogen from the air spaces in the soil directly into nitrates for use by the host plants. Such crops enrich the soil with nitrogen. All these processes constitute a sequence of events known as the **Nitrogen Cycle** (see Figure 37). This cycle will continue indefinitely provided that there is no interference by man.

In modern agriculture, crops are removed from the land where they are grown and human excreta is deposited elsewhere – sometimes in the sea. Thus, on agricultural land, artificial fertilizers are needed to replace the loss. These include ammonia itself, ammonium salts, nitrates, specially treated sewage, and urea. Urea, $CO(NH_2)_2$, the first organic compound to be synthesized, can be made from ammonia and carbon dioxide. Its nitrogen content by mass is very high – nearly 47%. In Figure 37, the dotted lines show the parts of the cycle contributed by man.

66   Phosphorus (Group V)

The chief source of phosphorus is calcium orthophosphate, $Ca_3(PO_4)_2$, which is found in N. Africa and elsewhere. This occurs as phosphatic rock, which is ground, mixed with silica and coke, and heated in an electric furnace:

$$2Ca_3(PO_4)_2(s) + 6SiO_2(s) + 10C(s) \longrightarrow 6CaSiO_3(s) + 10CO(g) + P_4(g)$$
$$\text{calcium silicate}$$

There are probably two stages to this reaction. First, the phosphate reacts with the silica to form calcium silicate and phosphorus(V) oxide:

$$2Ca_3(PO_4)_2 + 6SiO_2 \longrightarrow 6CaSiO_3 + P_4O_{10}$$

Then the oxide is reduced by the coke:

$$P_4O_{10} + 10C \longrightarrow 10CO(g) + P_4 \text{ (vapour)}$$

The vapour, consisting of tetratomic molecules, is condensed under cold water. As the white form of phosphorus has more intrinsic energy than the red, it is the white form that condenses. The element has few uses and most of it is burnt to form phosphorus(V) oxide from which the very important ortho-phosphoric acid and other acids can be obtained.

Phosphorus has two well-known allotropic forms (white and red). It is monotropic (Section 50) and the red form is the more stable. The two allotropes show remarkable physical and chemical differences (see Table 19).

Table 19

Properties	White (yellow) form	Red form
Physical		
Smell	Garlic-like	None
Crystalline form	Cubic	Rhombohedral
Rel. density (ord. temp.)	1.82	2.15–2.35
m.p.	44.25 °C	590 °C under pressure
b.p.	287 °C (760 mm)	Sublimes about 400 °C
Sp. ht.	0.20	0.17
Carbon disulphide	Soluble	Insoluble
Chemical		
Oxidation	Oxidises when exposed to the air	Does not
Ignition temp. in air	about 35 °C	About 260 °C
Heat of combustion	$-1550\,kJ\,mol^{-1}$	$-1505\,kJ\,mol^{-1}$
Hot alkali solution	gives phosphine gas	No action

- $1505\,kJ\,mol^{-1}$

To prove that both varieties are allotropes of one element, a known mass of each element can be burnt in excess oxygen. It is found that 1 g of each variety produces 2.29 g of phosphorus(V) oxide and nothing else.

The conversion of red to white and white to red can be carried out, on a small scale, in one simple apparatus (Figure 38).

Carbon dioxide (from a cylinder or Kipp's apparatus) is passed through a hard glass tube. A small amount of the red allotrope is placed in the tube a little way from the end where the inert gas enters and a small fragment

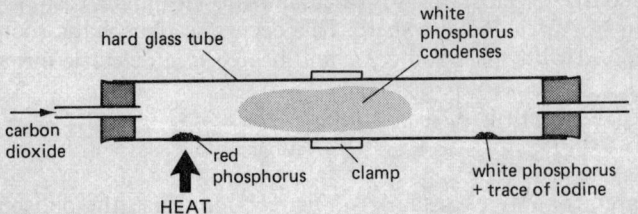

Figure 38 Small scale apparatus for the conversion of the allotropes of phosphorus

of white phosphorus (with a trace of iodine as catalyst) is placed near the other end of the tube. (White phosphorus is always cut under water holding the stick firmly with tongs.) The white phosphorus is warmed very gently and the conversion to the red form can be seen. The red variety is then heated more strongly. It sublimes and the vapour, carried by a slow current of carbon dioxide further along the tube, condenses as the white variety.

Industrially, red phosphorus is obtained by covering the white variety with a thin layer of water and heating it, in steel pots, under pressure for several hours at about 400 °C. Its molecular constitution is still unknown. The red form is much safer than the white because it does not oxidize when left in air. White phosphorus will ignite at 35 °C and when left in air the temperature rises as it oxidizes until it bursts into flame. It is always kept under water. The white variety is extremely toxic and great caution is necessary when handling it. A burn caused through careless handling does not heal as an ordinary burn does.

To prove that the white variety dissolves in carbon disulphide without chemical change, a very small piece can be dissolved in the cold organic solvent and the solution poured onto about three thicknesses of filter paper placed on a tripod. The solvent evaporates rapidly and leaves an invisible layer (probably in the P_4 molecular condition) on the paper. After a few seconds, the element catches fire spontaneously.

Because of the danger, white phosphorus is not used for making matches. Red phosphorus, however, is used in the mixture on the striking sides of a safety-match box. The top third of the pine match-stick is usually impregnated with paraffin wax and the head itself contains combustible material such as sulphur, oxidizing agents such as potassium chlorate or dichromate, with 'fillers' and 'binders'. A sulphide of antimony, Sb_2S_3, is usually used with red phosphorus on the striking surface and can also be in the match-head mixture.

All the common compounds of phosphorus are covalent and the valence of the element is 3 or 5. Phosphorus(V) chloride is well known but there is no parallel nitrogen(V) chloride although phosphorus is next below nitrogen in Group V. The electronic configuration of the nitrogen atom $(Z = 7)$ is 2,5. It cannot have more than eight electrons in its outer (second) shell and it achieves this in the ammonium ion, NH_4^+ (Section 10). Thus a compound NH_5 would seem to be impossible. On the other hand, the electronic configuration of a phosphorus atom $(Z=15)$ is 2,8,5. Its outer (third) shell can contain up to a maximum of eighteen electrons so PCl_5, which involves five pairs of electrons, is possible.

Binary phosphorus(III) compounds can be converted into phosphorus(V) by using excess of the particular reagent. If phosphorus burns in excess air or oxygen, P_4O_{10} is formed (a white solid). If the phosphorus is in excess, P_4O_6 is formed (both these oxides exist as dimers or double molecules). Similarly with the two chlorides PCl_5 and PCl_3. Phosphorus(V) chloride will dissociate on heating into phosphorus(III) chloride and chlorine (Section 19). On the other hand, the liquid phosphorus(III) chloride can be converted to the solid phosphorus(V) chloride by dripping the liquid from a tap funnel into a bottle through which dry chlorine is passed (the experiment should be performed in a fume chamber) (Figure 39):

$$PCl_3(l) + Cl_2(g) \rightleftharpoons PCl_5(s)$$

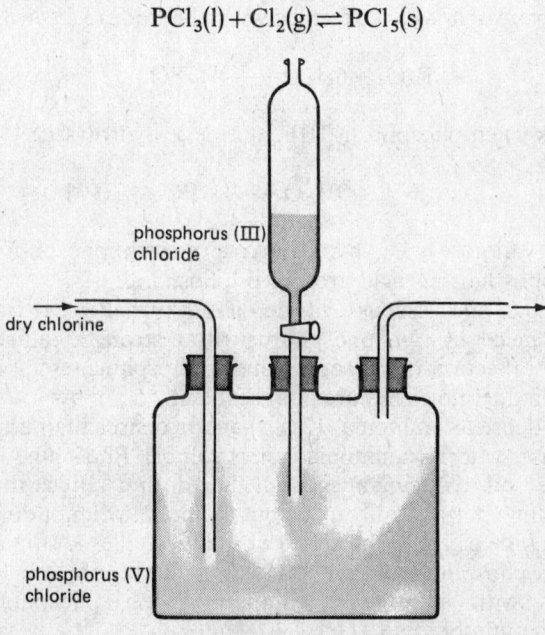

Figure 39

The oxide P_4O_{10} is made by burning white phosphorus in excess oxygen. It reacts with excess water to give orthophosphoric acid:

$$P_4O_{10} + 6H_2O \longrightarrow 4H_3PO_4$$

This is a strong tribasic (Section 33) acid which can be prepared in the laboratory (in a fume chamber) by the oxidation of red phosphorus (the action with white being too dangerous) with concentrated nitric acid. Torrents of brown fumes are evolved:

$$P + 5HNO_3 \longrightarrow H_3PO_4 + H_2O + 5NO_2$$

When carefully dehydrated the ortho- acid loses water in stages giving successively the 'pyro-' acid, the 'meta-' acid, and finally the phosphorus(V) oxide (the anhydride):

$$2H_3PO_4 - H_2O \longrightarrow H_4P_2O_7 \quad \text{(pyrophosphoric acid)}$$
$$(2)H_3PO_4 - (2)H_2O \longrightarrow (2)HPO_3 \quad \text{(metaphosphoric acid)}$$
$$2H_3PO_4 - 3H_2O \longrightarrow (P_2O_5) \quad \text{(phosphorus(V) oxide)}$$

All three acids have corresponding salts which have important uses, many connected with foodstuffs. Orthophosphates are used in rust proofing, manufacture of fertilizers, and the preparation of 'soft' drinks and jellies. Pyrophosphates are used in cheese processing, baking powders, and self-raising flour. Sodium metaphosphate polymer, $(NaPO_3)_n$, is an important water softener. Phosphates also help to stabilize colloids, as in emulsion paints and synthetic detergents, and metaphosphates are found in some tonics.

Phosphorus(III) oxide, also a white solid, is the anhydride of another acid called phosphorous acid which is formed when the oxide reacts with cold water:

$$P_4O_6 + 6H_2O \longrightarrow 4H_3PO_3$$

It also forms when phosphorus(III) chloride is hydrolysed:

$$PCl_3 + 3H_2O \longrightarrow H_3PO_3 + 3HCl$$

(Phosphorus(V) chloride, when hydrolysed, gives orthophosphoric acid.) Salts derived from phosphorous acid are called 'phosphites'.

Phosphorus has one gaseous hydride – phosphine, PH_3. It is obtained by warming a few pieces of *white* phosphorus with a strong alkali solution in an inert atmosphere (Section 75). In some respects phosphine resembles ammonia, NH_3; in others it does not. They are both covalent, colourless gases. Phosphine is just twice as dense as ammonia (17:8.5) and is denser than air. It is almost insoluble in water whereas ammonia is very soluble. Phosphine is only feebly basic. They are both reducing agents but they do not reduce the same compounds. Phosphine burns readily in air but ammonia will not do so unless the air is enriched with oxygen. The shape of the molecules is similar and phosphine can form phosphonium compounds (containing the ion PH_4^+) which are analogous with ammonium compounds, e.g. phosphonium iodide parallels ammonium iodide:

$$PH_3 + HI \longrightarrow PH_4^+I^-$$

67 Fertilizers

A good soil contains four main ingredients: clay, sand, lime, and plant food which is mainly organic and is known as 'humus'. Water cannot pass through clay but sand drains very quickly. A soil with too much clay may become waterlogged and with too much sand become parched. Lime (usually in the form of calcium hydroxide or calcium carbonate) keeps the soil 'sweet' by neutralizing excess acid.

The plant food must contain three important elements: potassium, nitrogen, and phosphorus. Traces of other elements, such as cobalt, copper, manganese, and zinc, are also needed. Nitrogen encourages leaf growth and is required particularly for grass, the brassica family – cabbages, brussels sprouts, etc. Phosphorus promotes root growth and potassium is important for fruit-bearing plants such as raspberries, strawberries, tomatoes, etc. When farmers had small mixed farms, ordinary farmyard manure provided humus containing all the essential ingredients. In modern times artificial fertilizers are necessary (Section 65). Humus, which also conserves moisture, can be supplemented, to some extent, by spent hops from breweries, garden compost-heaps, and peat (though the latter provides no nutrient). The three vital elements are provided artificially thus:

Potassium: Three soluble compounds the sulphate, K_2SO_4, the chloride, KCl, and the carbonate, K_2CO_3, come mainly from mineral deposits such as those in Stassfurt, Germany. Large deposits, which are now being exploited, have been found deep below ground in Yorkshire. Bonfire ash contains a useful amount of the carbonate.

Nitrogen: Provided by liquid ammonia, ammonium compounds, urea, nitrates, 'nitrolime' (Section 54), 'nitrochalk' (Section 63).

Phosphorus: Provided by 'basic slag', 'superphosphate', or 'triple super-phosphate'. Basic slag, a waste product from the steel industry (Section 71), is not very soluble, and is therefore slow-acting, and does not possess a very high phosphorus content. The common phosphorus compound, calcium orthophosphate (Section 66), is insoluble in water. Lawes, of Rothamsted, in 1842, discovered that if calcium orthophosphate is treated with 70% sulphuric acid it is converted to 'superphosphate' (actually calcium dihydrogen ortho-phosphate) which is soluble in water:

$$5Ca_3(PO_4)_2 + 11H_2SO_4 \longrightarrow 4Ca(H_2PO_4)_2 + 2H_3PO_4 + 11CaSO_4$$
'superphosphate'

This equation is not quantitative and can only indicate final products and approximate proportions. A drawback to the method is that it produces a large quantity of insoluble, relatively useless, calcium sulphate. A much better product results if the orthophosphate is treated with orthophosphoric acid itself. This gives the 'superphosphate' with no unwanted by-products:

$$Ca_3(PO_4)_2 + 4H_3PO_4 \longrightarrow 3Ca(H_2PO_4)_2$$

It is sold as 'triple superphosphate'.

68 Sulphur (Group VI), its Oxides, and Common Hydride

SULPHUR

Sulphur occurs in nature either as the element or in combination as sulphides and sulphates. It is next below oxygen in Group VI of the Periodic Classification and there are many resemblances between these two elements. Both oxygen gas and sulphur vapour are diatomic (Section 3). Sulphur will combine with nearly all the metals that oxygen combines with. Both elements show catenation (Section 52). When dilute acids react with oxides, water, H_2O, is formed; when they react with sulphides, hydrogen sulphide, H_2S, is formed. Hydrogen sulphide is a gas at ordinary temperatures whereas water is a liquid because oxygen causes hydrogen bonding (Section 11) but sulphur does not. Sulphur is less electronegative than oxygen. The atoms of sulphur have six valence electrons. Some of its divalent compounds are covalent; others electrovalent. By forming co-ordinate linkages sulphur can be tetravalent or hexavalent. Examples of tetravalence, such as sulphur dioxide, SO_2, are difficult to explain by simple valence theory. Oxygen seldom has a valence other than two, although it has a covalency of three in the oxonium ion H_3O^+ (Section 10).

Most of the world's sulphur comes from underground deposits. These were found in Louisiana and Texas, 160 m below ground, when prospectors were searching for oil. Owing to the difficult nature of the strata above, conventional mining methods were impossible. In 1891 the sulphur was successfully extracted by the **Frasch process.** Herman Frasch invented a method whereby no man is required to work underground. A steel tube, about 30 cm in diameter, is sunk into the sulphur bed and inside this is inserted the Frasch pump which consists of three concentric tubes (Figure 40). Down the outer tube, superheated water (under considerable pressure, to raise the temperature to about

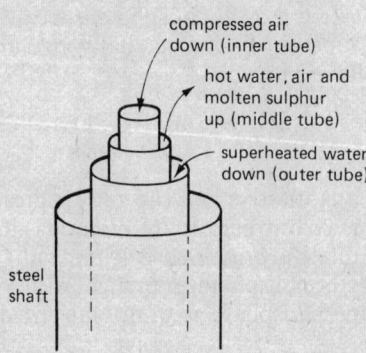

Figure 40 Frasch process for the extraction of sulphur from underground deposits

160 °C) is forced. This melts the sulphur (m.p. 119 °C) in the region round the bottom of the pump. Compressed air is then sent down the innermost pipe which is only about 4 cm in diameter. The air forms a froth with the sulphur and hot water, thus reducing the density of the mixture that has to be forced to the surface up the middle one of the three tubes. The sulphur that finally solidifies is 99.5% pure.

In volcanic areas, sulphur can still be recovered by melting the sulphur to separate it from the lava (mainly pumice and sulphur) and then distilling it in iron retorts (b.p. sulphur 445 °C). The inside of the iron retorts is attacked by the molten sulphur but a hard protective layer of iron(II) sulphide quickly forms.

If hydrogen sulphide gas is obtainable cheaply, pure sulphur can be extracted from it simply by burning it with a limited supply of air in specially designed kilns:

$$2H_2S + O_2 \longrightarrow 2H_2O + 2S$$

or by mixing the gas with sulphur dioxide gas:

$$SO_2 + 2H_2S \longrightarrow 2H_2O + 3S$$

For example, if sulphur has to be removed from petroleum it is extracted by way of hydrogen sulphide. Interest has also been taken in the conversion of sewage to methane and hydrogen sulphide by bacterial action. Likewise, in N. Africa, bacteria can produce sulphur from sulphates in the water, but the process is slow and may never be of commercial importance. At Lacq, S. France, the natural gas contains about 15% hydrogen sulphide and by passing it, with air, over an activated bauxite catalyst, sulphur is obtained.

Sulphur has two chief allotropic forms (Section 50): rhombic and monoclinic (Figure 41).

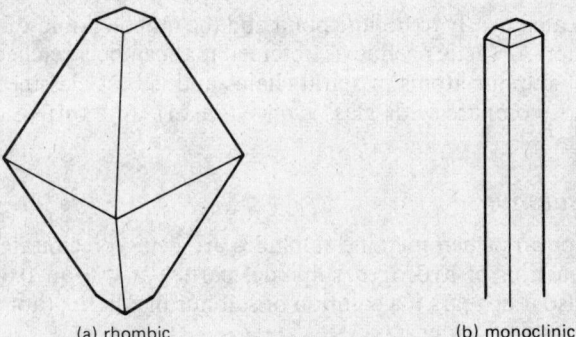

(a) rhombic   (b) monoclinic

Figure 41   Two allotropes of sulphur

Both varieties have molecules containing eight atoms, S_8, but the arrangement of the atoms is different. In the rhombic form the atoms are arranged in an eight-link ring, as shown in Figure 42.

Sulphur is enantiotropic with a transition temperature between these two forms of 95.6 °C. There are also amorphous forms, such as plastic sulphur, but these revert at room temperature to the stable rhombic form.

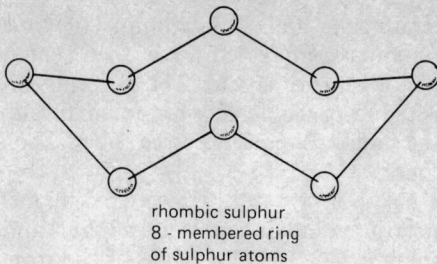

rhombic sulphur
8 - membered ring
of sulphur atoms

Figure 42

Rhombic or α-sulphur

Best made by crystallization from a solvent, usually carbon disulphide. Crushed sulphur is agitated for a few minutes with the solvent and then filtered. The filtrate is placed in a crystallizing dish and covered with a filter-paper in which a few pin holes have been made. Evaporation is thus slow enough to leave rhombic crystals of reasonable size.

Monoclinic or β-sulphur

(Also called 'prismatic' sulphur.)

Sulphur is melted in a small basin giving a temperature well above the transition temperature. On cooling, needle-like crystals can be seen forming across the surface of the molten element. When a crust has formed, two holes are made in it and the still molten sulphur inside is poured out. Breaking away the crust shows the monoclinic crystals clearly; but at room temperature the sulphur soon reverts to the rhombic form. The monoclinic variety can also be formed by crystallizing it from a boiling solvent whose boiling point is above the transition temperature, e.g. from toluene, C_7H_8, of b.p. 111 °C.

Plastic sulphur

Sulphur is heated nearly to boiling point and the mobile liquid quickly poured into cold water. A plastic product is obtained that can be stretched like elastic. It consists of sulphur atoms in spiral chains and is best classified as a super-cooled liquid (compare with glass in Section 61). It returns, if left, to the rhombic form.

Colloidal sulphur

This often appears when metallic sulphides are being precipitated from solution by the action of hydrogen sulphide, particularly if an oxidizing agent is present. Also it appears if a solution of sulphur in ethanol (not very soluble) is added to distilled water.

In the laboratory sulphur is usually seen as fairly-pure sticks (roll sulphur) or in finely crystalline form ('flowers of sulphur'). When heated in a test-tube, sulphur melts at the melting point of rhombic sulphur, 113 °C, because there has been no time for the transition. The melting point of monoclinic sulphur is about 6 °C higher. After melting there are successive changes in colour and viscosity, as heating continues, probably accompanying the breakdown of the molecules, $S_8 \rightarrow S_6 \rightarrow S_4 \rightarrow S_2$; the latter is the molecular state at the boiling point, 445 °C. At first, the molten liquid becomes darker in colour and very viscous; but when nearing the boiling point it becomes lighter in colour and much more mobile.

Important uses of sulphur include: manufacture of sulphuric acid, vulcanization of rubber, manufacture of gunpowder and safety matches, as a fungicide, formation of hydrogensulphites for paper making, and in organic chemistry for making dyes, sulpha-drugs, carbon disulphide, etc.

SULPHUR DIOXIDE, SO_2

Industrially this gas is obtained by burning sulphur in air or oxygen or as a by-product of the extraction of metals such as lead and zinc. In the laboratory it is obtained by warming a metal (usually copper) with concentrated sulphuric acid or by the action of a dilute acid on a sulphite (Section 75). It is a dangerous pollutant of the atmosphere as with moisture it forms sulphurous acid which then oxidizes to sulphuric acid. If breathed, therefore, lung damage soon results (as in the London 'smog'). As it is so soluble in water, its removal from industrial fumes should not be difficult.

With water it forms sulphurous acid, stronger than carbonic as an acid but weaker than the common mineral acids. It is dibasic (Section 33) and can form hydrogensulphites (bisulphites) or normal sulphites with alkali solutions:

$$SO_2 + H_2O \longrightarrow H_2SO_3 \quad \text{(sulphurous acid)}$$
$$H_2SO_3 + NaOH \longrightarrow NaHSO_3 \quad \text{(sodium hydrogensulphite)} + H_2O$$
$$H_2SO_3 + 2NaOH \longrightarrow Na_2SO_3 \quad \text{(sodium sulphite)} + 2H_2O$$

Sulphur dioxide is a powerful reducing agent and also a bleaching agent. As the latter it is less powerful, but also less destructive, than chlorine. It bleaches by reduction whereas chlorine bleaches by oxidation. The gas is usually identified by its smell (of burning sulphur), acid reaction, and its effects on potassium manganate(VII) solution and potassium dichromate solution. In the first of these, the purple colour disappears because of the reduction from manganese(VII) to manganese(II):

$$\underset{\text{purple}}{2MnO_4^-} + 5SO_2 + 2H_2O \longrightarrow \underset{\text{colourless}}{2Mn^{2+} + 5SO_4^{2-} + 4H^+}$$

In the case of potassium dichromate, the colour changes from orange to green as chromium(VI) is reduced to chromium(III):

$$\underset{\text{orange}}{Cr_2O_7^{2-}} + 3SO_2 + 2H^+ \longrightarrow \underset{\text{green}}{2Cr^{3+} + 3SO_4^{2-}} + H_2O$$

The reaction between sulphur dioxide and hydrogen sulphide gases is catalysed by water and in this case hydrogen sulphide reduces the sulphur dioxide:

$$2H_2S(g) + SO_2(g) \longrightarrow 2H_2O(l) + 3S(s)$$

Sulphur dioxide is easily liquefied (b.p. $-10\,°C$ at 1 atm.) by pressure alone and is supplied to laboratories in special metal containers. On releasing the pressure the gas is evolved.

Hot concentrated sulphuric acid will oxidize sulphur itself to sulphur dioxide:

$$S + 2H_2SO_4 \longrightarrow 3SO_2 + 2H_2O$$

The molecular formula of sulphur dioxide is established by burning sulphur in oxygen in an apparatus similar to that used for carbon dioxide (Section 49). If all measurements are made under the same physical conditions there is no change of volume when sulphur burns in oxygen. Therefore one molecule of sulphur dioxide must be formed from sulphur and one molecule of oxygen. The relative density of sulphur dioxide is 32 which gives a relative molecular mass of 64 corresponding to the formula SO_2.

$$\begin{aligned}
&\text{1 mole oxygen} + \text{sulphur} \longrightarrow \text{1 mole of the sulphur gas} \\
&\quad\ \ O_2 \qquad + \quad xS \quad \longrightarrow \qquad\qquad S_xO_2
\end{aligned}$$

If $x = 1$ the relative molecular mass is 64.

The important uses of sulphur dioxide include bleaching, as a fungicide and pesticide, manufacture of sulphites and sulphuric acid.

SULPHUR (VI) OXIDE, SO₃ (SULPHUR TRIOXIDE)

Sulphur dioxide and oxygen do not normally react with each other, but if the two gases are passed over a heated catalyst, a white silky solid can be isolated which is sulphur trioxide, SO_3. On a laboratory scale the two gases are passed through concentrated sulphuric acid (so that the rate of flow can be observed and regulated) with the oxygen in excess, to make sure that most of the dioxide is converted. The mixed gases pass over platinized asbestos heated in a hard-glass tube (Figure 43). The gas and smoke formed are then passed through a U-tube kept cold in an ice/water mixture. A white silky solid deposits at the bottom of the U-tube. (The temperature must not be too low because some dioxide might become liquefied.)

$$2SO_2 + O_2 \rightleftharpoons 2SO_3$$

Sulphur trioxide melts at 17 °C and vaporizes easily (b.p. 45 °C). On re-

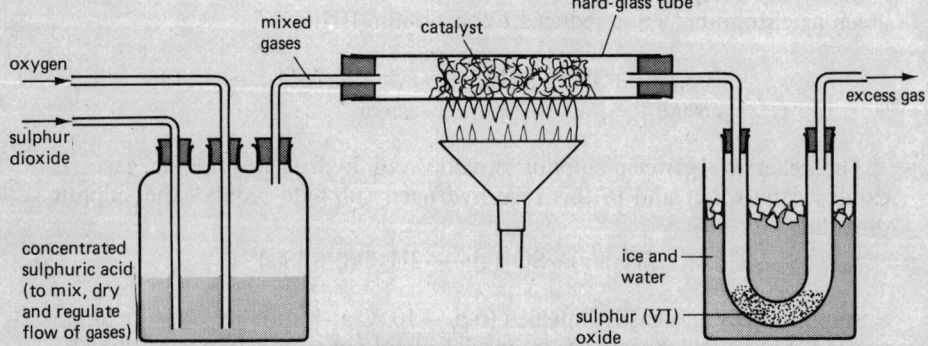

Figure 43   Preparation of sulphur(VI) oxide

moving the U-tube a little distilled water can be dripped very cautiously onto the white solid. There is a violent reaction:

$$SO_3 + H_2O \longrightarrow H_2SO_4$$

If the solution is divided into two parts one part can be tested with litmus solution and the other part for sulphate ion (Section 77). Sulphurous acid gives a white precipitate with barium chloride solution of barium sulphite but this is soluble in dilute hydrochloric acid; so a white precipitate in the presence of excess hydrochloric acid shows sulphate ion.

HYDROGEN SULPHIDE, H₂S (SULPHURETTED HYDROGEN)

This very poisonous evil-smelling gas is usually prepared in the laboratory by the action of moderately concentrated hydrochloric acid on iron(II) sulphide (Section 75). The latter is seldom free from iron and if a purer gas is required antimony(III) sulphide, Sb_2S_3, is used instead of the iron compound:

$$FeS + 2HCl \longrightarrow FeCl_2 + H_2S$$
or $$Sb_2S_3 + 6HCl \longrightarrow 2SbCl_3 + 3H_2S$$

Hydrogen sulphide is a fairly soluble gas in water and the solution is a weak acid. It is dibasic (Section 33) so hydrogensulphides and normal sulphides can be prepared. Concentrated sulphuric acid cannot be used to dry the gas because it would be reduced:

$$H_2S(g) + H_2SO_4(l) \longrightarrow SO_2(g) + 2H_2O(l) + S(s)$$

Anhydrous calcium chloride is usually chosen.

The gas burns (blue flame) in a limited supply of air to produce sulphur and steam but with excess of air or oxygen, sulphur dioxide and steam are formed:

$$2H_2S + 3O_2 \longrightarrow 2SO_2 + 2H_2O$$

Apart from its smell of rotten eggs, it is usually recognized by its power of turning lead ethanoate paper black through the formation of lead(II) sulphide. It is an important reducing agent and has the same effect on potassium manganate(VII) solution and potassium dichromate solution that sulphur dioxide has *except* that sulphur is nearly always precipitated when hydrogen sulphide reduces a compound. Its hydrogen is oxidized to water and sulphur is freed. Sulphur dioxide does *not* blacken lead ethanoate paper. If hydrogen sulphide is passed through dilute nitric acid a white precipitate of sulphur is seen.

Hydrogen sulphide can be synthesized from its elements, but the reaction is reversible and incomplete:

$$H_2 + S \rightleftharpoons H_2S$$

If pure hydrogen is passed through boiling sulphur, hydrogen sulphide can be detected in the issuing gas.

The gas is used in analytical chemistry for recognizing many metal ions by ion aggregation (Section 18). The colour of the precipitate (ppt) will often indicate the presence of a particular metal ion.

In the presence of excess hydrochloric acid:

Pb^{2+} ions give *black ppt* of the sulphide. So also do Hg^{2+}, Ag^+, and Cu^{2+} ions.

Cd^{2+} gives *yellow ppt*, As^{3+} *dull yellow*, Sb^{3+} *orange*.

In alkaline solution of aqueous ammonia:

Zn^{2+} ions give *white ppt* of the sulphide.

Mn^{2+} gives *pink ppt*.

Ni^{2+} gives *black ppt*.

For example, with lead:

$$Pb(NO_3)_2 + H_2S \longrightarrow PbS + 2HNO_3$$

or, more simply $Pb^{2+} + S^{2-} \longrightarrow PbS$

Note: If hydrogen sulphide is passed through an iron(III) salt solution there is *no* precipitate of sulphide, only a sulphur precipitate as the gas reduces iron(III) to iron(II) the colour changing from yellow/brown to pale green (Section 71):

$$2Fe^{3+}(aq) + H_2S(g) \longrightarrow 2Fe^{2+}(aq) + 2H^+(aq) + S(s)$$

Sulphides of the Group I metals are colourless, ionic crystalline compounds that are quite soluble in water. Sodium sulphide, Na_2S is very deliquescent and is used for removing hair from hides. Group I sulphides are always alkaline in solution because, being salts of a weak acid, they are hydrolysed to some extent (Section 20).

69   Sulphuric acid

If sulphur dioxide is passed into water, sulphurous acid, H_2SO_3, is formed. Left in the air, this will slowly oxidize to sulphuric acid. This straightforward method of making sulphuric acid, H_2SO_4, has never been an economic proposition but, basically, methods of manufacturing the acid have used the same three ingredients: air (or oxygen), sulphur dioxide, and water. The old Chamber Process (now nearly if not quite extinct) combined the three with the help of oxides of nitrogen as the catalyst. It produced acid, not very pure, of about 70% concentration which was excellent for preparing the fertilizer 'super-phosphate' (Section 67). It has not been able to compete with the Contact Process.

Contact Process

This method, invented about a century ago, first puts together the sulphur dioxide and oxygen and then adds the water. It produces very pure acid. The sulphur dioxide comes from the burning of sulphur or as a by-product of the extraction of lead and zinc.

The sulphur dioxide, purified if necessary, is mixed with excess of air and passed over a heated catalyst at 400–450 °C. The platinum catalyst that was formerly used has been replaced by vanadium(V) oxide, V_2O_5, because it is more efficient, cheaper, and less easily 'poisoned' by impurities. The sulphur trioxide fumes are passed not into water, because of the violence of the reaction, but into fairly concentrated sulphuric acid. This absorbs the trioxide continuously and quietly, eventually forming 'fuming' sulphuric acid (also called 'oleum'). This can be diluted to give acid of any desired concentration. The chain of events can be formulated:

$$S(s) + O_2(g) \longrightarrow SO_2(g)$$
$$2SO_2(g) + O_2(g) \rightleftharpoons 2SO_3(g) \qquad \Delta H = -189\,kJ$$
$$SO_3(g) + H_2SO_4(l) \longrightarrow H_2S_2O_7(l)\ (oleum)$$
$$H_2S_2O_7 + H_2O \longrightarrow 2H_2SO_4$$

All the above reactions are exothermic. Considering the second reaction, raising the temperature should, by Le Chatelier's Principle, send the reaction from right to left; but if the temperature is too low the reaction is too slow. 400–450 °C is chosen as a reasonable 'optimum' working temperature. Again, as in this reaction three volumes of mixed gas give two volumes of trioxide vapour, higher pressures should increase the yield. However, as the final yield is quite good, the expense of extra pressure is not justified.

The acid is an oily liquid of relatively high density ($1.83\,g\,cm^{-3}$ at 20 °C) and b.p. of 338 °C. When heated, there is decomposition before the boiling point is reached and white fumes can be seen. When diluting, the cold acid must be added cautiously to cold water with continuous stirring, because of the great heat evolved. The strongly exothermic reaction between acid and water is explained by the formation of hydrates. Concentrated sulphuric acid is a powerful dehydrating agent (illustrated by its charring of paper and wood and the removal of water from ethanol or sucrose). It is a strong dibasic acid (Section 33) and the hot acid can act as a mild oxidizing agent (illustrated by its reaction with carbon or sulphur – each oxidized to the dioxide). The acid must always be treated with the greatest caution.

Sulphuric acid is immensely important to industry and there is hardly a manufacturing process that does not need it at some stage. Its main uses include the preparation of sulphates, other acids (as sulphuric acid is almost non-volatile), fertilizers, dyes, rayon, soap and glycerol, explosives, storage batteries, purification of oils.

70   Transition elements

Element number 20 of the Periodic Classification, calcium, has the electronic configuration 2,8,8,2. It has eight electrons in the third shell and two in the fourth. The third shell, however, is capable of accommodating a maximum of eighteen electrons. For the elements of atomic numbers 21–30 inclusive, as element succeeds element the extra electron goes not into the fourth shell but into the third until, in the case of the last one, zinc, the third shell is completely full. For element 31 (gallium) and succeeding ones, additional electrons continue to fill the fourth shell. In the first transition series there are ten electron spaces to fill in the third shell, i.e. five orbitals that can take two electrons each, as illustrated by the boxes in Figure 44.

It will be noticed that no electron pairs with another until all the orbitals each have one electron. So, apart from zinc, all these elements have at least one

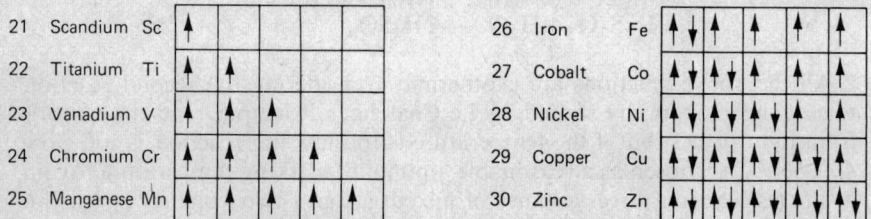

Figure 44

unpaired electron in the third shell. Such unpaired electrons can be 'promoted', by gaining extra energy, to the fourth shell and act as additional valence electrons. Because the atoms of each element have two valence electrons in the fourth shell, each element will have a minimum valence of 2 (though copper is an exception here) and a maximum valence of $2+n$, where n represents the number of unpaired electrons in the third shell. One would expect, therefore, that manganese could have possible valencies of 2, 3, 4, 5, 6, and 7. All these are known and most have been mentioned in earlier sections, e.g. manganese(II) (formerly called 'manganous' salts, such as $MnSO_4$); manganese(III) (formerly called 'manganic' salts, such as the fluoride MnF_3); manganese(IV) (MnO_2); manganese VI (manganates, $MnO_4{}^{2-}$); manganese(VII) (permanganates, $MnO_4{}^-$).

In the Periodic Classification there are three series of transition elements all of which are metals. Two of the first series will be studied briefly – iron and copper. Briefer mention is made of silver in the second series and of platinum and gold in the third series. The study of the structure and properties of the elements of the second and third series is more complicated than that connected with the first series and cannot be considered here.

Possession of unpaired electrons in the penultimate shell gives to the transition elements some unusual properties. The chief of these are:

Variable valence: Briefly explained above.

Coloured ions: The possession of unpaired electrons or transfer of electrons from one energy level to another is always associated with colour or change of colour. Some common examples are: Mn(II), pink; Fe(II), green; Fe(III), brown; Co(II) (hydrated ion) pink; Ni(II), green; Copper(II) (hydrated ion), blue. Zinc(II) is not associated with colour and its common salts are colourless. Zinc chromate, $ZnCrO_4$, is yellow but this colour comes from the chromate ion CrO_4^{2-}.

Catalytic power: Many transition elements have catalytic powers that are explained by the formation of intermediate compounds – made possible by the activity of the unpaired electrons. Iron (Haber Process) and nickel (hydrogenation) are examples.

Magnetic properties: Substances that are attracted by a magnet are said to be *paramagnetic*. Iron, cobalt, and nickel are very paramagnetic and these are known as 'ferromagnetic' metals. The phenomenon is associated with the presence of unpaired electrons in the atoms.

Complex ion formation: Transition elements readily form complex ions (Section 37), e.g. iron and copper (Sections 71 and 72).

As the penultimate shell of a zinc atom has its full complement of electrons, zinc is the last of the first transition series and the element does not possess the properties associated with unpaired electrons. Its valence is always 2, its common salts are not coloured, it is not used as a catalyst, and it has no paramagnetic properties. Its chemistry resembles that of the two elements whose electron configurations are analogous: magnesium, 2,8,2; calcium, 2,8,8,2; zinc, 2,8,18,2.

71   Transition element: iron

Iron can have valencies ranging from 2 to 6 as a maximum (Section 70). Three of these are well known: iron(II) ('ferrous'); iron(III) ('ferric'); and iron(VI) ('ferrates' – which are analogous with chromates and have the ion FeO_4^{2-}). Compounds of iron(IV) and iron(V) have now also been obtained.

Extraction of iron from its ores

Iron occurs in nature in many forms and in many places. The chief ores are haematite (iron(III) oxide, Fe_2O_3) so named because of its red colour; limonite (hydrated iron(III) oxide, 'bog iron ore'); siderite (iron(II) carbonate, $FeCO_3$); magnetite (the compound oxide, Fe_3O_4); and iron pyrites(FeS_2). Such ores are found in the U.S.A., Sweden, Britain, Germany, Spain, Russia, Canada, Brazil, and China, but only the first four of the ores listed are commonly used for the extraction of the metal.

The crushed ore is mixed with coke and ignited on a sintering machine (Section 44, Figure 20) which provides material of the right size for the blast furnace. The sintered ore is then mixed with coke and limestone and poured into the furnace. Air, preheated to about 700 °C, is blown into the base of the furnace through the tuyères (pronounced 'twyers'). At the hottest part of the molten zone the temperature reaches about 1800 °C (Figure 45).

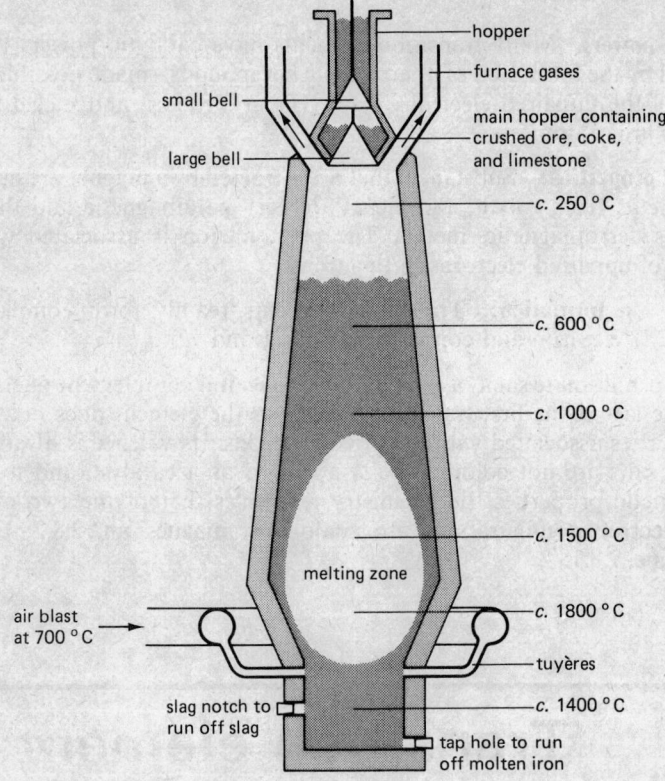

Figure 45   Blast furnace

The chief reactions that occur are:

(1) At temperatures above 700 °C oxygen from the air combines with carbon to form the monoxide:

$$2C + O_2 \longrightarrow 2CO$$

(2) Assuming that haematite is used, the haematite is reduced by the carbon monoxide, probably in stages: $Fe_2O_3 \longrightarrow Fe_3O_4 \longrightarrow FeO \longrightarrow Fe$. Overall the reaction is exothermic:

$$Fe_2O_3 + 3CO \longrightarrow 2Fe + 3CO_2$$

(3) The limestone is decomposed into calcium oxide and carbon dioxide:

$$CaCO_3 \longrightarrow CaO + CO_2$$

(4) The calcium oxide (basic) unites with the unwanted silica (acidic) in the ore, forming a slag of calcium silicate:

$$CaO + SiO_2 \longrightarrow CaSiO_3$$

The slag forms a molten layer on top of the molten iron at the base of the furnace and the carbon dioxide formed in (2) and (3) is quickly converted to more carbon monoxide:

$$CO_2 + C \longrightarrow 2CO$$

The double cone device on the hopper ensures that the furnace can be recharged without loss of furnace gas. This gas leaves the top of the furnace at about 250 °C and its volume composition is approximately 60% nitrogen, 25% carbon monoxide, 10–15% carbon dioxide, with small quantities of other gases. It carries with it much dust but, when this has been removed, the gas can be used to pre-heat the air blown in at the base. Some of the molten iron is run into moulds to form crude 'pig-iron' but usually it is transferred immediately to furnaces of a different type for the manufacture of steel. There is little demand for pure iron.

The iron from the blast furnace may contain as much as 4% carbon (by mass), 2% silicon, and about 1% each of manganese and phosphorus, depending on the nature and quality of the ore used. The carbon exists in the iron partly in solid solution and partly combined as carbide. Objects made from cast iron are relatively cheap but brittle.

Steel making

The essential requirements for steel making are ridding the iron of unwanted elements (chiefly by oxidation), the adjusting of the carbon content, and the addition of small quantities of other elements to obtain the type of steel required. Very low carbon steel ('wrought iron') is relatively soft and malleable. It contains less than 0.15% carbon. Mild steels contain 0.15–0.25% carbon, medium-carbon steels 0.25–0.50%, and high-carbon steels 0.5–1.4%.

OPEN-HEARTH PROCESS: This has been used for over a century but is gradually becoming obsolete. The furnace is a reverberatory one, i.e. the charge is not heated directly by the fuel but by the hot gases, produced from the fuel, passing over the charge spread out on the 'bed' of the furnace. The charge consists of molten iron from the blast furnace mixed with rusty scrap iron, extra ore, and limestone. The process is slow and takes from four to twenty hours to complete. Oxygen (from the additional oxide and rust from the scrap) slowly oxidizes the unwanted elements, some of which pass off as gases (e.g. CO_2 and SO_2) and some (such as P_4O_{10}) are absorbed in the slag. The fuel could be natural gas, coke oven and blast-furnace gas mixture, or 'atomized' oil. At each end of the furnace bed are heat regenerators of chequer brickwork (called Cowper stoves) and the mixture of fuel gas and air is passed over the bed first in one direction and then in the other. The incoming gas is therefore preheated by the hot brickwork. The process can be hastened by 'lancing' the molten charge with oxygen and the two advantages that the method still has over its rivals are that (1) it can deal with several hundred tonnes at one time and (2) it makes excellent use of scrap.

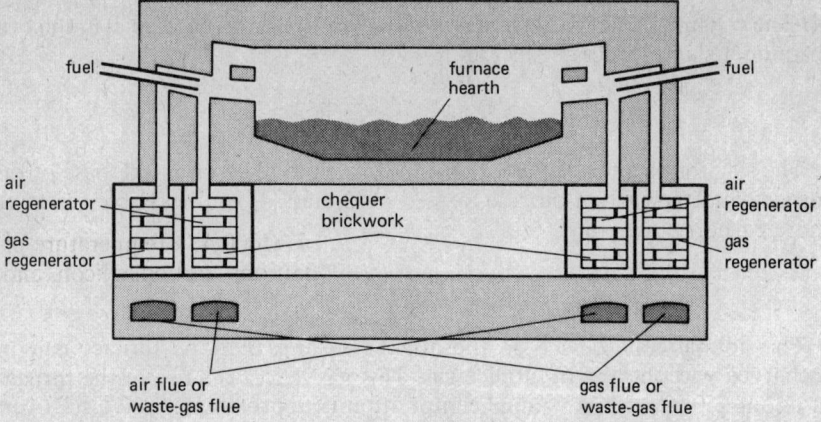

Figure 46 Open-hearth furnace

BESSEMER CONVERTER: Inverted by Henry Bessemer in 1856, this furnace deals with smaller quantities but the process is quicker and does not need additional fuel. The converter can be charged with 30–40 tonnes of molten crude iron whilst in the horizontal position (Figure 47(a)) and then it can be swung into the vertical position whilst a blast of air under pressure is blown through (Figure 47(b)).

The oxidation of the impurities produces much heat and a temperature of 1650 °C is quickly reached (m.p. pure iron 1535 °C). If the air blast is enriched with oxygen, the 'blow' can be completed in about ten minutes and an oxygen/steam mixture can reduce the time still further. Oxygen alone would produce too fierce a reaction and if air is used some nitrogen can be absorbed by the molten metal. Carbon and sulphur form their dioxides and, if phosphorus is known to be present, the converter is previously lined with calcined dolomite (magnesium and calcium oxides); this basic lining combines with the phosphorus(V) oxide to give phosphate – subsequently ground and sold as 'basic slag' (Section 67).

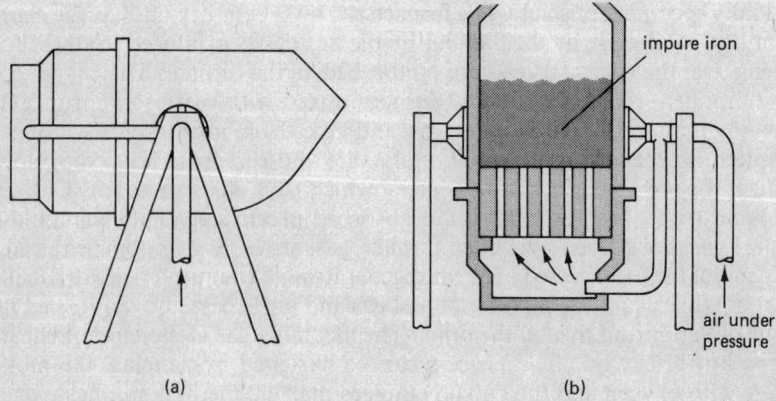

(a)   (b)

Figure 47 Bessemer converter (a) being charged in the horizontal position and (b) converter in vertical position

Modifications of this process, such as the L.D. and the Kaldo (Swedish), blow oxygen onto the surface of the molten iron.

ELECTRIC FURNACES: These are only economical where electric power is cheap. They cannot cope with large quantities of crude iron, but they give a much finer control of temperature and addition of alloying materials. The process can be carried out under oxidizing or reducing conditions. The furnaces can be either of the arc or high-frequency type, the former being the more important. With a voltage of 200 V and current of 30 000 A, a temperature of 2400 °C can be reached and, under oxidizing conditions, carbon, silicon, and phosphorus are oxidized and taken up by the furnace lining to form an easily-removable slag.

BISRA CONTINUOUS PROCESS: Methods outlined above suffer from the drawback that they are 'batch' methods; that is to say the process has to be stopped at regular intervals to remove products and then the process recommenced. The British Iron and Steel Research Association (BISRA) have developed a process in which molten iron, straight from the blast furnace, is 'atomized' into a spray by high-velocity jets of oxygen, lime and fluxes being continuously added to form a slag which is run off. The purified iron is caught in ladles. Alloying elements can then be added.

Many types of steel are made and often the name indicates the element that has been added: manganese steel (7–20% by mass Mn) is very hard; molybdenum steel (2% Mo) does not soften at high temperatures; nickel steel (5% nickel) has high wear-resistance; silicon steel (5% Si) is used for the cores of electromagnets; tungsten steel (up to 20% W) is used for high-speed tools. There are hundreds of specifications for stainless steel but the common one is the '18/8' variety – 18% chromium and 8% nickel. It has a thin surface film of the very inert chromium(III) oxide, Cr_2O_3, that protects it from corrosion. It resists the action of air, moisture, alkalis, and acids, though concentrated hydrochloric acid can damage it.

Rusting of iron and steel

Rusting requires the presence of both air and water. Bright iron nails kept in a desiccator remain bright indefinitely – they have air but no water. If distilled water, that has been well boiled, is placed hot in a test-tube, bright nails added, and the surface sealed with paraffin wax, the nails will not rust – they have water but no air. (When added to the hot water, the wax melts, floats on the surface, and, when cool, seals it.) Nails left in an open dish, partly covered by water, rust in a very short time – they have both air and water. That acid gases, such as carbon dioxide or sulphur dioxide, also affect rusting can be shown by pushing a large bright nail through the cork of a small flask containing a little of the solution of a strong alkali. If the whole nail is thoroughly wetted and the apparatus left for a day or two, the top of the nail (exposed to the air) rusts quickly, but the bottom of the nail – surrounded by moist acid-free air – does not.

The composition of rust varies widely and its colour can be any shade from yellow to almost black. Iron(III) oxide and hydroxide are the important products, but the composition depends on the conditions prevailing when the rusting takes place. It is now known that *rusting is an electrolytic process and the presence of an electrolyte is essential.* The rusting process is greatly promoted if the water contains carbon dioxide or sulphur dioxide, which produce ions. Iron near the sea rusts very rapidly because salt-spray provides the electrolyte. It has been shown that if there are two points on an iron surface with an electrolyte between them and one is under more oxidizing conditions than the other, corrosion takes place at the point that is *least* oxidized. This explains why, when painted iron is damaged and some of the metal is exposed to the damp atmosphere, the parts remote from the air corrode rather than those exposed to the air.

The surface of iron or steel is very irregular. If there is a film of water on the metal, small humps will be exposed to the air (cathodic areas) whilst anodic areas will be below the surface. Hydrogen ions travel to the cathodic areas and hydroxide ions to the anodic areas. Iron atoms lose electrons to form iron(II) ions which produce a precipitate of iron(II) hydroxide with the hydroxide ions. This is subsequently oxidized to iron(III) hydroxide. At the cathodic areas hydrogen ions may form very small quantities of hydrogen.

So much money is continually wasted by the corrosion of iron and steel that many ways of preventing it have been devised:

(1) *Covering the surface with oil or grease:*   Efficient but temporary.
(2) *Painting:*   Effective as long as the paint surface is not broken.
(3) *Coating with a 'sacrificial' metal:*   e.g. galvanizing (Section 45).
(4) *Electroplating:*   e.g. with tin. Efficient but costly. Chromium gives an attractive finish, but the surface is porous and rusting is not prevented. Very good plating has first a layer of copper, then nickel, and finally chromium.
(5) *Alloying with other metals:*   e.g. stainless steel, as described above.
(6) *Chemical treatment of the surface* to produce an inert protective layer: Could be an oxide film or a phosphate film (obtained by pickling in ortho-phosphoric acid).
(7) *Use of inhibitors:*   In cases where water has to remain for some time in contact with iron, compounds can be added to the water which inhibit the formation of rust. Such additives include chromates and sodium benzoate. These act as negative catalysts for the rusting process.

If iron or steel has to remain below ground, e.g. a pipe line, not only are water, air, and various ions present but bacteria as well. Even stainless steel does not resist these. The remedy is to sheathe the pipe in a plastic material. On a smaller scale, rods of zinc or magnesium can be buried near the pipe and connected to it by wires. The non-ferrous metal is then corroded sacrificially.

Compounds of iron

Iron has two common oxides, the dull-red iron(III) oxide and the black compound oxide, Fe_3O_4, which acts as a mixture of iron(II) and iron(III) oxides. Iron(II) oxide itself, FeO, is so easily oxidized that, on exposure to air, it quickly becomes iron(III) oxide, Fe_2O_3.

Iron(III) oxide is easily obtained in the laboratory by heating iron(II) sulphate ('green vitriol'). The alchemists used the reaction as a source of sulphuric acid:

$$2(FeSO_4,7H_2O) \longrightarrow 14H_2O + SO_3 + SO_2 + Fe_2O_3$$

Black iron oxide, Fe_3O_4, is produced whenever iron is heated strongly in air or in steam, as in Lavoisier's 'gun-barrel' experiment (Section 26):

$$3Fe + 2O_2 \longrightarrow Fe_3O_4$$
$$3Fe + 4H_2O \rightleftharpoons Fe_3O_4 + 4H_2 \qquad \text{(Section 48)}$$

The hydroxides of iron(II) (green) and iron(III) (brown), both nearly insoluble in water, can be precipitated by ion aggregation from the appropriate solutions, e.g. iron(II) from iron(II) sulphate solution and iron(III) from iron(III) chloride solution. Ammonium or sodium hydroxide solution can be used as the precipitating agent:

$$Fe^{2+} + 2OH^- \longrightarrow Fe(OH)_2 \quad \text{(dull-green precipitate)}$$
$$Fe^{3+} + 3OH^- \longrightarrow Fe(OH)_3 \quad \text{(brown precipitate)}$$

The two chlorides of iron are well known. Iron(II) chloride can be obtained by passing dry hydrogen chloride over heated iron wires and iron(III) chloride by passing dry chlorine over the red-hot iron:

$$Fe + 2HCl(g) \longrightarrow FeCl_2(s) + H_2(g)$$
$$2Fe + 3Cl_2(g) \longrightarrow 2FeCl_3(s)$$

In the case of the iron(III) chloride, a guard tube is fitted to the receiving flask to protect it from any water vapour that would hydrolyse the black anhydrous iron(III) chloride (Figure 48). The common brown iron(III)chloride used in the laboratory is the hexahydrate, $FeCl_3, 6H_2O$.

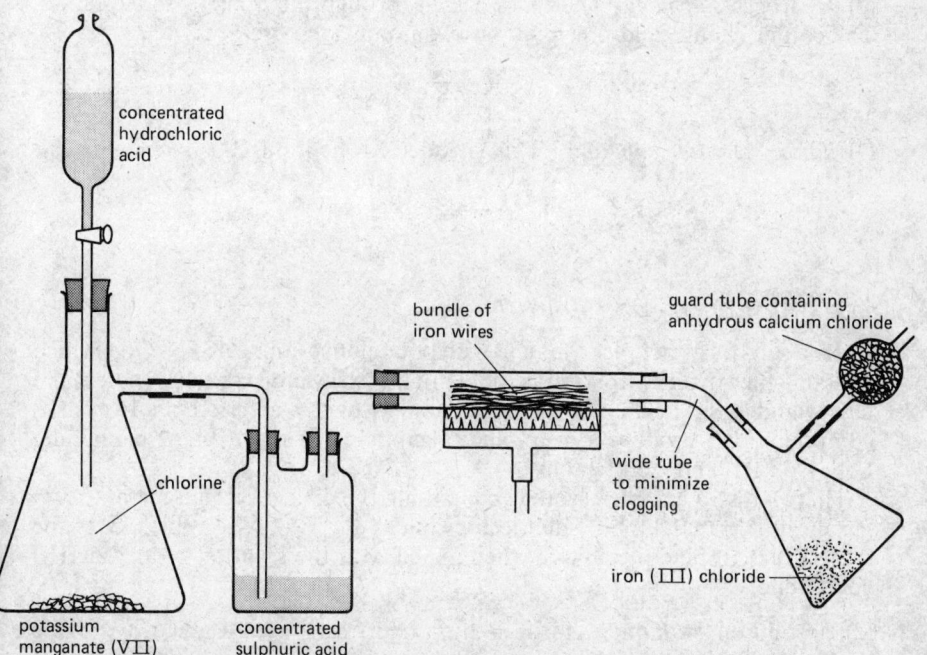

concentrated hydrochloric acid

bundle of iron wires

guard tube containing anhydrous calcium chloride

chlorine

wide tube to minimize clogging

iron (III) chloride

potassium manganate (VII)

concentrated sulphuric acid

Figure 48 Preparation of iron(III) chloride

Iron(III) sulphate is usually encountered as 'iron alum' which is the double salt with ammonium sulphate, $(NH_4)_2SO_4,Fe_2(SO_4)_3,24H_2O$ (Section 42).

Most iron(II) salts oxidize slowly in the air to the iron(III) state. If a solution of iron(II) sulphate (green) is heated in air, the solution begins to turn brown. Almost without exception, iron(II) salts have a green colour and iron(III) salts a brown one.

TO CHANGE IRON(II) IN AQUEOUS SOLUTION TO IRON(III):
Many oxidizing agents will effect the change including nitric acid, hydrogen peroxide, chlorine, acidified manganate(VII) or dichromate solutions. The colour change green → brown is seen at once (as the oxidizing solution is colourless).

TO CHANGE IRON(III) IN AQUEOUS SOLUTION TO IRON(II):
Reducing agents that easily effect this change include 'nascent' hydrogen, hydrogen sulphide, and tin(II) chloride solution.

(1) *'Nascent' hydrogen:* If ordinary molecular hydrogen gas is bubbled through an iron(III) salt solution, nothing happens, but if zinc and hydrochloric acid are added to the solution to be reduced and the *hydrogen evolved in the solution*, reduction does take place. This is why the reducing agent was referred to as 'nascent' hydrogen (from the Latin 'nascor' meaning 'I am born (or produced)'). It is likely that in the presence of hydrochloric acid, zinc supplies electrons for the reduction of iron(III) ions and the reaction is a straightforward electron transfer:

$$Zn + 2Fe^{3+} \longrightarrow Zn^{2+} + 2Fe^{2+}$$

(2) *Hydrogen sulphide:* The gas is passed through the iron(III) salt solution. The colour change and precipitation of sulphur are observed:

$$2Fe^{3+} + H_2S \longrightarrow 2Fe^{2+} + 2H^+ + S(s)$$

(3) *Tin(II) chloride solution:* This is added to the iron(III) solution.

$$2Fe^{3+} + Sn^{2+} \longrightarrow 2Fe^{2+} + Sn(IV)$$

Tests for iron(II) and iron(III)

(1) Iron salts form complex cyanides with potassium cyanide, KCN. One, which contains the trivalent iron ion, is potassium hexacyanoferrate(III) (potassium ferricyanide), $K_3[Fe(CN)_6]$ which, when mixed with a solution of any iron(II) compound will produce a deep-blue precipitate ('Prussian blue') containing both iron(III) and iron(II) ions.

There is also a complex cyanide containing the divalent iron ion, potassium hexacyanoferrate(II) (potassium ferrocyanide), $K_4[Fe(CN)_6]$, which gives the same Prussian blue precipitate when mixed with the solution of an iron(III) compound.

(2) A solution of ammonium or sodium hydroxide gives a green hydroxide precipitate from an iron(II) solution but a brown precipitate from an iron(III) solution.

(3) A solution containing thiocyanate ions, CNS^- (colourless) when added to the solution of an iron(III) salt gives a deep-red colouration – *not* a precipitate – due to complex formation. The test is very sensitive.

Table 20

Test	Reagent	Iron(II)	Iron(III)
(1)	$K_3[Fe(CN)_6]$	*Dark-blue ppt*	Dark-brown solution
	$K_4[Fe(CN)_6]$	White ppt	*Dark-blue ppt*
(2)	NH_4OH or NaOH	*Green ppt* $(Fe(OH)_2)$	*Brown ppt* $(Fe(OH)_3)$
(3)	KCNS or NH_4CNS	No colour	*Deep-red colour*

72 Transition element: copper

The copper atom has two electrons in the fourth shell and only one unpaired electron in the third. Therefore one would expect copper to possess two possible valencies 2 or 3. Strangely, its common valencies are 1 and 2. Valence 3 is very uncommon and the monovalent copper compounds result when one of the electrons of the fourth shell fills the remaining gap in the third – this, under certain conditions, being the more stable arrangement. With divalent copper, the two electrons of the fourth shell are the valence electrons. Hydrated divalent copper ions, $Cu(H_2O)_4^{2+}$, have a blue colour in aqueous solution but the monovalent copper compounds are mainly insoluble in water and colourless. In elementary chemistry, copper(II) compounds are more important than those of copper(I).

Extraction of copper

The element occurs widely in nature, mostly in combination. The chief ores are sulphides, with lesser quantities of basic carbonate (malachite) and basic sulphate (brochantite). The one used for the extraction of copper is chalcopyrite (copper pyrites), $CuFeS_2$. The metallurgy of copper is complicated by the necessary removal of the large quantity of iron associated with the copper. When this has been done, the semi-purified ore is placed in a converter, rather like the one used in steel making, and an air blast is blown through the molten metal for about three hours. The temperature reaches about 1300 °C and at this high temperatue copper(I) sulphide, Cu_2S, is the stable sulphide. In the converter the copper(I) sulphide is partially converted into copper(I) oxide, Cu_2O, and then these two interact to produce fairly pure 'blister copper' – the blisters being caused by the escape of gas as the metal cools and solidifies:

$$2Cu_2S + 3O_2 \longrightarrow 2Cu_2O + 2SO_2$$
$$2Cu_2O + Cu_2S \longrightarrow 6Cu + SO_2$$

Blister copper can be refined either thermally or electrolytically. The former is good enough for the metal used for making pipes, boilers, condensers, etc., but the latter is always used to produce the very pure metal needed for electrical work. Any impurity lessens the metal's conductivity and most of the world's copper is now electrolytically refined.

THERMAL REFINING: Molten blister copper is poured onto the bed of a reverberatory furnace at about 1200 °C and air, or oxygen, is blown through it to oxidize the unwanted elements. Sulphur burns off as the dioxide and the more electropositive elements present, such as zinc and iron, form oxides that can be skimmed from the surface. To prevent the copper itself from being oxidized, during the later stages poles of green wood are introduced so that the reducing gases expelled from the charring wood prevent such oxidation.

ELECTROLYTIC PROCESS: The blister copper becomes the anodes of electrolytic cells that have thin sheets of pure copper as cathodes. The electrolyte is copper(II) sulphate solution (containing about 4% copper by mass) with 1.5M sulphuric acid and a little glue to aid the deposition. The electrolyte slowly circulates at about 55 °C. Deposition continues for about two weeks:

At the cathode $(-)$ $\qquad$ $Cu^{2+} + 2e^- \longrightarrow Cu$
At the anode $(+)$ $\qquad$ $Cu - 2e^- \longrightarrow Cu^{2+}$

There is a steady transference of copper from anode to cathode and the concentration of the copper(II) solution does not change. At the anodes are deposited small quantities of valuable metals such as gold, silver, and platinum, which can be recovered. Other metals, e.g. iron and nickel, go into solution but under the prevailing conditions are not deposited.

Copper(I) compounds

These are not so common nor so important as the copper(II) compounds, but two of them are of interest in elementary chemistry – the oxide, Cu_2O, and the chloride, CuCl. Copper(I) oxide is dull-red in colour. (Oxide colours are unpredictable and bear little relationship to the colours of ions in aqueous solution.) It is formed when an aldehyde or aldehydic sugar is warmed with Fehling's solution (Sections 56 and 58). Fehling's solution has a rich purple colour and contains copper(II) complex ions which are reduced to a precipitate of copper(I) oxide. The reaction can be used for testing for the presence of glucose in human urine.

Copper(I) chloride, CuCl, forms when copper, copper(II) oxide, and concentrated hydrochloric acid are warmed together. Without the presence of extra copper the basic copper(II) oxide would give copper(II) chloride and water. With extra copper and the solution boiled gently for a few minutes, copper(I) chloride is formed:

$$CuO + Cu + 2HCl \longrightarrow 2CuCl + H_2O$$

If the hot solution is decanted into a beaker containing air-free distilled water, a white precipitate of copper(I) chloride is formed. If the mixture is stirred and filtered immediately, the white chloride is left on the filter paper. This residue soon begins to turn blue/green indicating oxidation to the copper(II) state. If, immediately after filtering, a concentrated solution of ammonia is poured through the filter the white solid dissolves to form a deep-blue solution containing complex ions. This solution of ammoniacal copper(I) chloride solution is used as an absorbing agent for carbon monoxide gas.

Copper(II) compounds

As some of these have been described in earlier sections, only a brief summary of the important ones is given here. The black oxide, CuO, is formed only superficially when air or oxygen is passed over red-hot copper. To obtain the oxide in a fairly pure state, pyrolysis of the hydroxide, basic carbonate or nitrate is possible.

Being low in the activity series, copper is not attacked when cold, by air, water, alkalis or non-oxidizing acids. Steam has no effect on copper which makes the metal particularly useful for the construction of hot-water systems, boilers, condensers, etc. It reacts with hot concentrated sulphuric acid to give the sulphate and other products; also very readily with nitric acid, even cold and dilute, to give the nitrate. Insoluble copper(II) compounds are therefore best obtained by ion aggregation using a nitrate or a sulphate solution. In this way the hydroxide (light-blue in colour) and the basic carbonate (light-green) can be obtained using the appropriate precipitating agents. Copper(II) salts in solution have a characteristic and unique reaction with a fairly concentrated ammonia solution. When the ammonia is first added, the light-blue precipitate of copper(II) hydroxide is seen. On adding more ammonia, with shaking, the precipitate disappears and a deep-blue solution containing the complex ions $[Cu(NH_3)_4]^{2+}$ is obtained.

The uses of metallic copper are chiefly connected with its excellent conductivity of heat and electricity and its non-reactivity with water or steam. They include the manufacture of telegraph and telephone wires, electric cables, electrical machinery, water or steam pipes, roofing material, cooking utensils, beaten ware (because it is very malleable). The green patina that forms on the surface of copper-clad roofs is due to the action of oxygen, carbon dioxide, and water on the metal which produces basic copper(II) carbonate. In industrial areas (where there may be sulphur dioxide in the atmosphere) some basic copper(II) sulphate may also be present.

A very large number of useful alloys are derived from copper including the following. (Composition given in percentages by mass.)

Muntz metal (Yellow metal): 40 zinc. Resists corrosion by sea water
Various brasses: 20–35 zinc
German silver (nickel silver): 10–30 nickel and a little zinc (used for EPNS (electroplated nickel silver) and latch keys)
Tin bronze: 5–8 tin
Aluminium bronze: 3–10 aluminium
Bell metal: 20 tin
Gun metal: $8\frac{1}{2}$–10 tin, 2–4 zinc
Constantan: 40 nickel (used for wire of resistance coils)
Monel: 68 nickel with about 1 each of iron and manganese. Has a high m.p. and excellent resistance to chemical corrosion.

73 Transition elements: silver, platinum, gold

These elements are sometimes known as 'noble metals' because of their value and resistance to corrosion. Silver is a member of the second transition series of the Periodic Classification (immediately under copper). Platinum is next to gold and they are members of the third series.

SILVER

Silver, like copper, can have a valence of one and this applies to all its common compounds. Apart from the nitrate, all these compounds are insoluble in water and, therefore, silver nitrate solution can cause the precipitation of many negative ions by ion aggregation.

The element occurs native and also as silver(I) sulphide, Ag_2S. More than half of the silver extracted comes from the desilverization of the ores of copper and lead. From the sulphide, this compound is soluble in a dilute solution of sodium cyanide by forming a complex ion. From this solution the metal is displaced by adding a more electropositive metal such as aluminium or zinc:

$$2Ag^+ + Zn \longrightarrow Zn^{2+} + 2Ag$$

'Standard' silver is 92.5% silver by mass, the rest being copper that has been added to harden it. This alloy was used as silver coinage in this country until 1920 when the percentage of silver was lowered to 50, the rest being nickel and copper. In 1946 silver coinage was replaced by 'cupronickel' (copper 75%, nickel 25%).

Silver is even lower in the activity series than copper and it resists the action of dilute acids, air, and water. Like copper, it is attacked by hot concentrated sulphuric acid and by nitric acid. It is also blackened by the action of hydrogen sulphide which forms silver(I) sulphide. Silver is sometimes coated with a thin layer of gold to prevent such tarnishing – 'silver gilt'.

Silver(I) nitrate, $AgNO_3$, is used to precipitate halides and thus to help in their identification (Section 77). Many other negative ions are precipitated by silver nitrate solution but these are all soluble in dilute nitric acid (e.g. the carbonate). Hence the silver nitrate test for halogens is always performed in the presence of excess dilute nitric acid. Silver nitrate is obtained by dissolving silver in dilute nitric acid and it crystallizes in colourless rhombic form. It is a caustic substance (the old name 'lunar caustic' still survives). If a solution is spilt onto the skin and not immediately washed away, a black stain appears because organic matter reduces it leaving silver as a black residue. Silver bromide, $AgBr$, is immensely important for the preparation of photographic film. Any silver halide left in the air begins to darken as light starts its decomposition – a well-known photochemical reaction (Section 46). To make this reaction rapid, silver bromide is dispersed colloidally over pure gelatin and this mixture coats the film. Because of photochemical decomposition, silver nitrate solution is always stored in dark bottles.

PLATINUM

This is extracted in small quantities from the nickel ores found at Sudbury, Ontario. Its relative density is nearly twice that of lead (21.5:11.3). Possible valencies are 1, 2, 3, 4, and 6. The metal is not attacked by air, water, or any acid except the 'aqua regia' mixture (see below). It is attacked by fused strong alkali and, in this respect, is not as resistant as nickel. The pure metal is used for making resistance thermometers and thermocouples. Hardened with small quantities of other transition metals it is used for jewellery, small pieces of chemical apparatus, and for the 'spinnerets' used for drawing threads of man-made fibres. It is an important catalyst. Having much the same coefficient of expansion as glass, platinum wire can be inserted into softened glass without the danger of cracking when the glass cools, e.g. platinum wires for flame tests.

GOLD

This valuable metal is mined in S. Africa, very deep below ground, and it may need one hundred tonnes of ore to yield 500 g of gold. The extraction from the crushed ore involves a cyanide process very similar to that used for silver. Some gold is recovered during the electrolytic refining of copper.

The common valencies of gold are 1 and 3. The metal has a relative density of 19.3 at 20 °C – much denser than lead but not as dense as platinum. Gold is very malleable and can be beaten into leaf that is translucent and almost transparent. Gold leaf is used for gilding other materials but there are now many cheap substitutes for this. Pure gold is 24 carat so that 9-carat gold is only about one-third gold, the rest being mainly copper which hardens and cheapens the metal without much alteration of colour. 18-carat gold is much used for jewellery, wedding rings, etc., but purer metal than this is too soft for most uses. Gold imitations include the Victorian 'Pinchbeck' ('poor man's gold') which was first used by Pinchbeck and is copper with 7–11% zinc. Another is 'Dutch metal' (copper with 20% zinc) which in leaf form looks like gold leaf but cannot be beaten so thinly.

Gold is not affected by air, water or any common acids except the 'aqua regia' mixture (hydrochloric acid with a little nitric acid – both concentrated). The acids interact producing some free chlorine as well as nitrosyl chloride, NOCl:

$$4H^+ + NO_3^- + 3Cl^- \longrightarrow NOCl + Cl_2 + 2H_2O$$

Chlorine reacts with every metal and the gold is eventually converted into gold(III) chloride, $AuCl_3$. Gold, like platinum, is attacked by fused strong alkali.

Colloidal gold is used in the preparation of good-quality ruby-coloured glass.

When a manufacturing company has to choose a site for its chemical plant many considerations have to be borne in mind. These include: ready availability of raw materials; supply of reasonably cheap power; easy transport for raw materials and products; disposal of waste; effect of effluent fluids on the neighbourhood. Too often, in the past, the last factor has been totally or partially ignored. It has been customary to site chemical factories close to large rivers because the latter provide easy transport, water for cooling systems, and are convenient for the dumping of waste liquor. In earlier days, this dumping has led to the fouling of rivers so that the normal ecology has been disrupted and fish are unable to survive. Pesticides, herbicides, etc., applied to the land also find their way into the rivers. During this century, people have gradually become aware of the dangers, and rivers such as the Thames and Humber are steadily being restored to normal – though there is still much to be done. With modern facilities, sewage and waste of all kinds can be converted into useful fertilizers, etc., and the liquid waste purged of unpleasant material; but the equipment for these processes is costly, and progress is slower than it ought to be. Those who live in or near the industrial areas of cities such as London, Manchester, and Birmingham know that there is still a long way to go. The siting of factories to produce such things as sulphuric acid, carbon black, dangerous organic compounds, and any type of nuclear plant presents special difficulties.

Pollution of the atmosphere

Air pollution, as defined by the World Health Organization, 'occurs when one or several air pollutants are present in such amounts and for such a long period in the outside air that they are harmful to humans, animals, plants or property, contribute to damage or may impair the well-being or use of property to a measurable degree.'

In the mid-nineteenth century, disaster was caused by those who started to manufacture washing soda by the now obsolete Leblanc process, letting dangerous gases such as hydrogen chloride and hydrogen sulphide escape into the surrounding countryside. Public outcry produced the first legislation to stem pollution – the Alkali Act of 1864. In our own times the cities of Los Angeles and San Francisco (both in California) have been bedevilled by 'smog' (a portmanteau word for conditions of smoke with fog). In London, in the years 1952 and 1956, smog caused many deaths through acute diseases of the respiratory tracts.

The most dangerous conditions occur in cities when (1) there is persistent heavy fog or (2) when there is a long hot sunny spell without rain. In the first case (as in the London smog disasters) soluble acid gases such as sulphur dioxide and nitrogen dioxide dissolve in water droplets to produce dangerous acids and these are not dispersed. It is, however, unlikely that such bad conditions will recur, because the Clean Air Act of 1956 has removed most of the smoke from the atmosphere of cities. Particles of soot provide nuclei on which, during fog, droplets form. In the second case, dangerous acid gases are not removed by rain and various photochemical reactions can go on undisturbed (as in the summer of 1976). Reaction between oxygen and nitrogen dioxide, particularly in the presence of very small quantities of unsaturated hydrocarbons (also from car-exhaust emission), can produce ozone – which is toxic even in small concentrations to living organisms.

The chief gases that now pollute our atmosphere are carbon monoxide, nitrogen dioxide, ozone, sulphur dioxide, and gases derived from lead (as well as dust of lead compounds). The chief sources of these are domestic heating appliances, car exhausts (from both petrol and diesel engines), and factory chimneys.

CARBON MONOXIDE, CO (Section 49): This is formed mainly by the incomplete combustion of carbon and hydrocarbons in stoves, engines etc. It is a deadly poison but small quantities in the air soon become oxidized to the dioxide.

NITROGEN DIOXIDE, NO$_2$ (Section 62): This escapes in small quantities into the air mainly from car exhausts. Explosions of petrol, or diesel fuel, with air in internal-combustion engines take place at a high enough temperature to produce a little nitrogen monoxide, NO, from the gases of the air, and this monoxide forms the dioxide as soon as it escapes into the open air. It is very soluble in water and is normally removed by rain. In long hot sunny spells, it can accumulate and act photochemically with oxygen to produce small quantities of ozone.

OZONE, O$_3$ (Section 30): Formed as explained above. Its concentration in the atmosphere approached unacceptable levels during the hot summer of 1976.

SULPHUR DIOXIDE, SO$_2$ (Section 68): The most widely distributed of the polluting gases. It is produced from domestic heating appliances using oil or solid fuel that contains small quantities of sulphur, from the exhaust fumes of motor cars, and from chemical industry, e.g. the smelting of sulphide ores, manufacture of sulphuric acid, from sulphites used in paper making, in bleaching and preserving processes. The gas is a strong irritant and with water forms sulphurous acid then, by oxidation, sulphuric acid. It can be removed from industrial gases by wet-scrubbing with aqueous alkaline solutions or by adsorption on carbon in the form of a special type of coke – which is reasonably cheap.

GASEOUS LEAD COMPOUNDS (Section 44): These tetraethyl lead compounds are formed from the 'anti-knock' additives used in petrol. Research has not so far found leadless substitutes that are as efficient. (Dust of lead compounds can also get into the atmosphere from furnaces producing lead and factories making storage batteries, paints, glass, etc.). Lead acts as a cumulative poison and the level of lead pollution in cities has to be very carefully watched.

75   Common gases and their preparation

General Notes on Information Given

R.M.M.   Relative Molecular Mass

R.D.   Relative Density (relative to hydrogen). If the density relative to air is required, divide the R.D. figure by 14.4 which is the density of air relative to hydrogen.

Solubility   refers to solubility in water. Any figures quoted are for s.t.p. and give volumes of gas per unit volume of water.

'*Dilute acid*'   Where this is specified, either hydrochloric (about 4M) or sulphuric acid (about 2M) will serve; *not* nitric which is an oxidizing acid.

HYDROGEN, H_2

R.M.M. $=2$ R.D. $=1$ Colourless. Odourless. Insoluble. Neutral. Flammable.

Laboratory Preparation:
Zinc with dilute acid.
$$Zn(s) + 2H^+(aq) \longrightarrow Zn^{2+}(aq) + H_2(g)$$

Kipp's apparatus can be used (Section 49, Figure 26(b)).
Gas dried through any common agent. Collected by upward delivery.
Industrial Preparation: Catalytic steam/hydrocarbon reforming process,

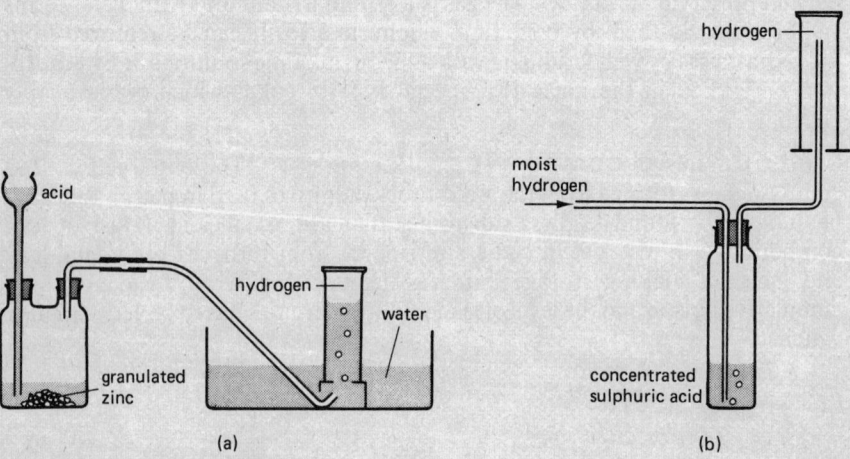

Figure 49   (a) Preparation of moist hydrogen.   (b) Collecting dried hydrogen gas

using methane or the light naphtha fraction from petroleum distillation. Catalyst is nickel mixed with a refractory cement. With methane:

$$CH_4(g) + 2H_2O(g) \rightleftharpoons 4H_2(g) + CO_2(g)$$

As the reaction, left to right, is endothermic and reversible (Sections 46 and 48) it can be assisted by high temperature (1000 °C). Carbon dioxide is removed by washing with water under pressure.

Hydrogen is also obtained as a by-product of the electrolysis of brine (Section 38) and from water gas (Section 49).

GASES THAT ARE COMPOUNDS OF CARBON
(Group IV)

Carbon Dioxide, CO_2

R.M.M. = 44   R.D. = 22   Colourless. Odourless. Somewhat soluble (1.7:1) giving a weakly acid solution.

Laboratory Preparation: Marble with dilute hydrochloric acid. Sulphuric acid cannot be used in this case because calcium sulphate is insoluble.

$$CaCO_3(s) + 2HCl(aq) \longrightarrow CaCl_2(aq) + CO_2(g) + H_2O(l)$$

Kipp's apparatus can be used (Section 49, Figure 26(b)).
Gas dried through concentrated sulphuric acid. Collected by downward delivery.
Industrial Preparation: By-product of fermentation processes (Section 59), lime kilns, from hydrocarbons as above. Combustion of carbon itself.

Carbon Monoxide, CO

R.M.M. = 28   R.D. = 14   Colourless. Odourless. Very poisonous (about same density as air). Insoluble. Flammable.

Laboratory Preparation: Dripping concentrated sulphuric acid onto warm methanoic acid in order to dehydrate it (Figure 50).

$$H.COOH(l) - H_2O(l) \longrightarrow CO(g)$$

Can also be obtained from the dioxide (Section 49, Figure 28).

Industrial Preparation: Has been used as a fuel (with nitrogen) in producer gas, and with hydrogen in water gas (Section 49).

Methane, CH_4

R.M.M. = 16   R.D. = 8   Colourless. Odourless. Insoluble. Neutral. Flammable.

Laboratory Preparation: Heating fused sodium ethanoate with soda-lime (Figure 51). The ethanoate is fused to make sure that it is free from water. Soda-lime is obtained by slaking quicklime with sodium hydroxide solution.

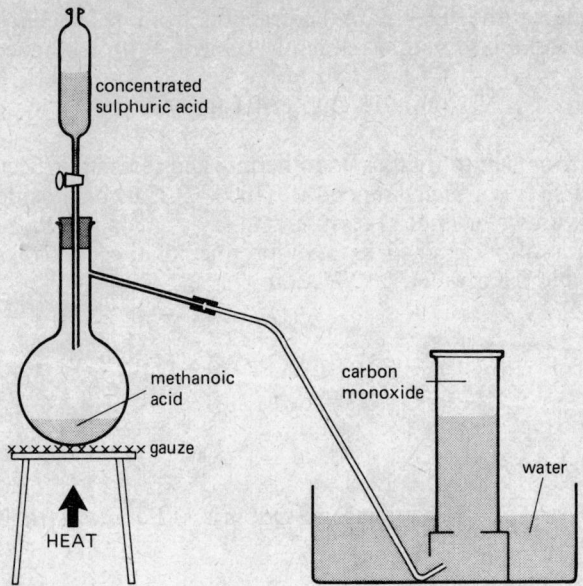

Figure 50   Preparation of carbon monoxide (to be done in a fume cupboard)

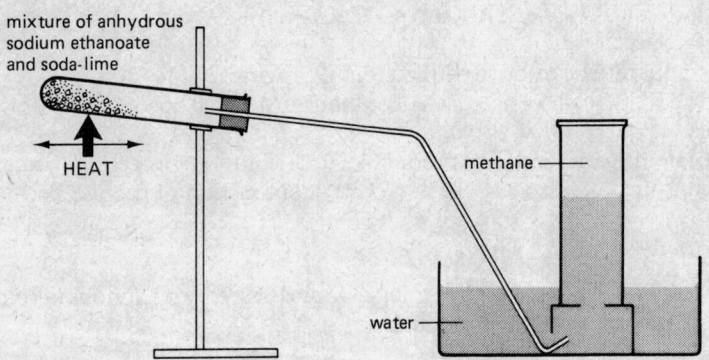

Figure 51   Preparation of methane

It is a mixture of calcium and sodium hydroxides, solid, and easier to handle than sodium hydroxide. It is not deliquescent.

$$CH_3.COONa(s) + NaOH(s) \longrightarrow Na_2CO_3(s) + CH_4(g)$$

Dried through concentrated sulphuric acid. Collected by upward delivery.

Industry: Occurs as natural gas.

Ethene (Ethylene), C_2H_4

R.M.M. $= 28$ R.D. $= 14$ Colourless. Odourless. Insoluble. Neutral. Flammable.

Laboratory Preparation: The dehydration of ethanol. This can be done by warming with an excess of concentrated sulphuric acid or more safely, on a

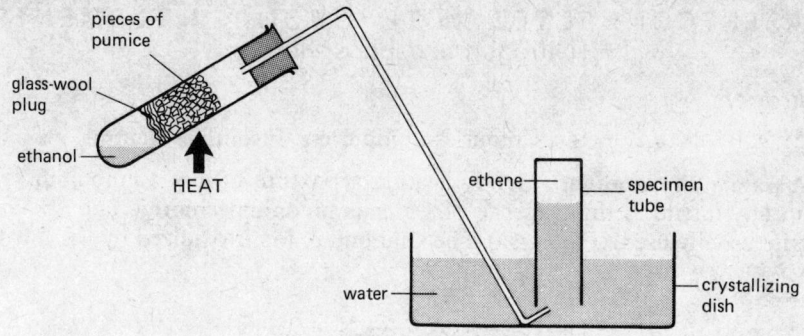

Figure 52 Small-scale preparation of ethene

small scale, by passing the vapour over a catalyst such as pumice (Figure 52).

$$C_2H_5OH(l) - H_2O(l) \longrightarrow C_2H_4(g)$$

Collected over water.

Industrial Preparation: Catalytic cracking of saturated hydrocarbons (Section 54).

Ethyne (Acetylene), C₂H₂

R.M.M. = 26   R.D. = 13   Colourless. Sweetish smell when pure but normally has a strong garlic-like odour. Insoluble. Flammable.

Laboratory Preparation: Dripping water onto calcium carbide (Figure 53).

$$CaC_2(s) + 2H_2O(l) \longrightarrow Ca(OH)_2(s) + C_2H_2(g)$$

Collected over water.

Industrial Preparation: As above. Also by the breakdown of natural gas under special catalytic conditions using also a little oxygen.

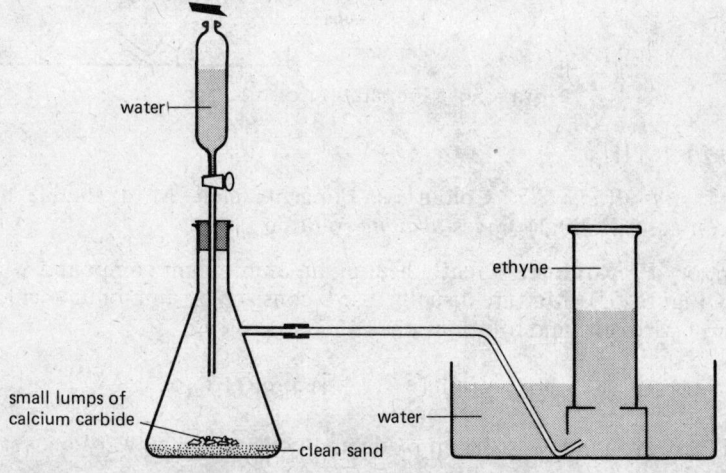

Figure 53   Preparation of ethyne

GASES CONNECTED WITH GROUP V ELEMENTS
(Nitrogen and phosphorus)

Nitrogen, N$_2$

R.M.M. = 28   R.D. = 14   Colourless. Odourless. Insoluble. Neutral.

Laboratory Preparation: Gently heating a mixture of any ammonium salt with any nitrite in the presence of water (ammonium chloride and sodium nitrite usually used) (Figure 54). The ammonium ion is oxidized by the nitrite ion.

$$NH_4^+ + NO_2^- \longrightarrow N_2 + 2H_2O$$

Can also be obtained, fairly pure, from the air (Section 30, Figure 10). Collected over water.

Industrial Preparation: Fractional distillation of liquefied air.

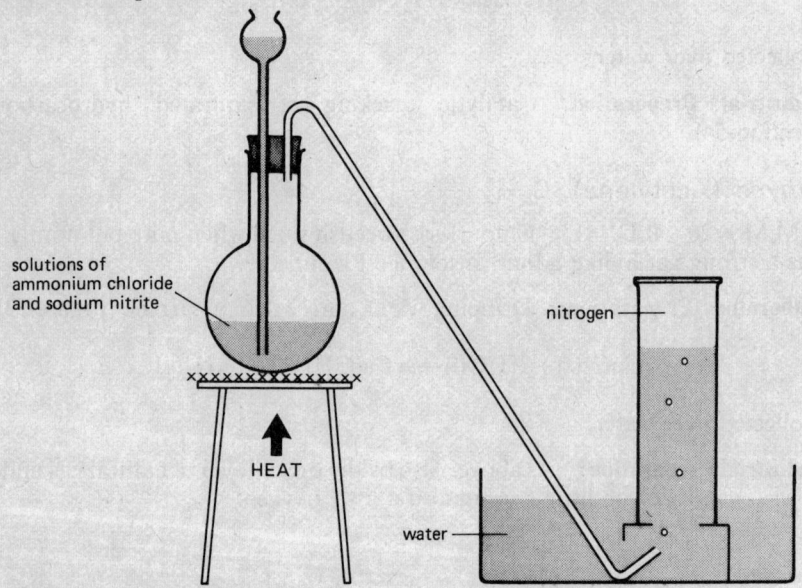

solutions of
ammonium chloride
and sodium nitrite

nitrogen

HEAT

water

Figure 54   Preparation of nitrogen

Ammonia, NH$_3$

R.M.M. = 17   R.D. = 8.5   Colourless. Pungent smell. Most soluble of the common gases (1300:1). Gives *alkaline* solution.

Laboratory Preparation: Gently heating an ammonium compound with an alkali (Figure 55). Mixture usually used consists of ammonium chloride, calcium hydroxide, and a little water.

$$NH_4^+ + OH^- \longrightarrow NH_3(g) + H_2O(l)$$

Dried through lumps of calcium oxide and collected by upward delivery.

Industrial Preparation: Haber Process (Sections 48 and 63).

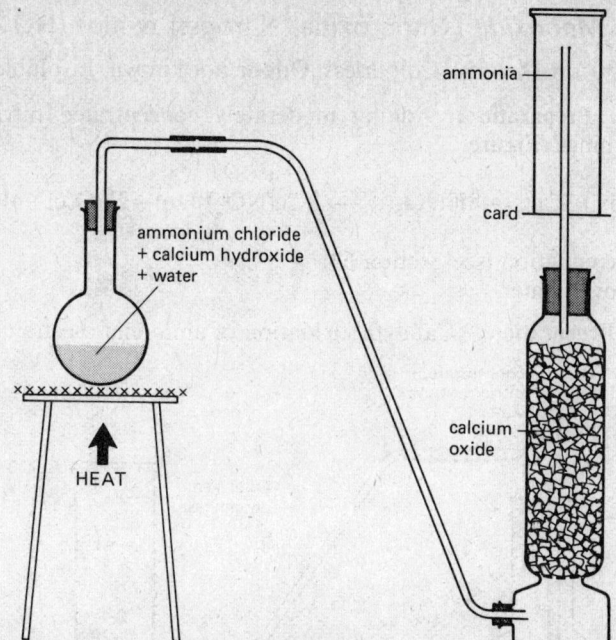

Figure 55   Preparation of ammonia

Dinitrogen Oxide (Nitrous oxide), N₂O

R.M.M. = 44   R.D. = 22   Colourless. Sweetish sickly smell. Somewhat soluble (1:1). Neutral.

Laboratory Preparation: Heating ammonium nitrate (Figure 56). Heating must be stopped while at least half the nitrate remains because, if the temperature rises too much, there is the risk of an explosion.

$$NH_4NO_3(s) \longrightarrow N_2O(g) + 2H_2O(g)$$

Collected over hot water.
Industrial Preparation: As above.

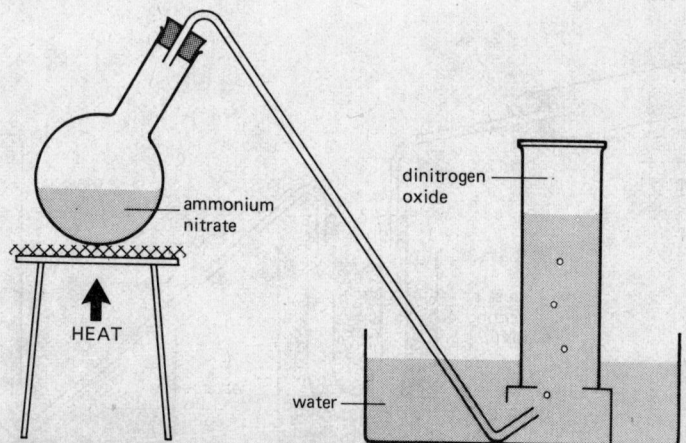

Figure 56   Preparation of dinitrogen oxide

Nitrogen Monoxide (Nitric oxide, Nitrogen oxide), NO

R.M.M. = 30   R.D. = 15   Colourless. Odour not known. Insoluble. Neutral.

Laboratory Preparation: Adding moderately concentrated nitric acid to copper turnings (Figure 57).

$$(\text{approx.}) \quad 3Cu(s) + 8HNO_3(l) \longrightarrow 3Cu(NO_3)_2(aq) + 2NO(g) + 4H_2O(l)$$

No reliable equation (see Section 64).
Collected over water.

Industrial Preparation: Catalytic oxidation of ammonia (Section 64).

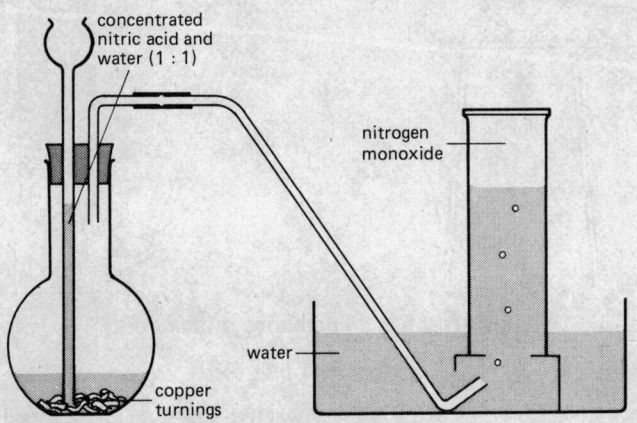

Figure 57   Preparation of nitrogen monoxide

Dinitrogen Tetroxide, N₂O₄

R.M.M. = 92   R.D. is usually between 23 and 46. Pure N_2O_4 is liquid and boils at 22 °C. It is almost colourless. On heating it dissociates into brown nitrogen dioxide, NO_2. When evolved in a test-tube the gas is some mixture of

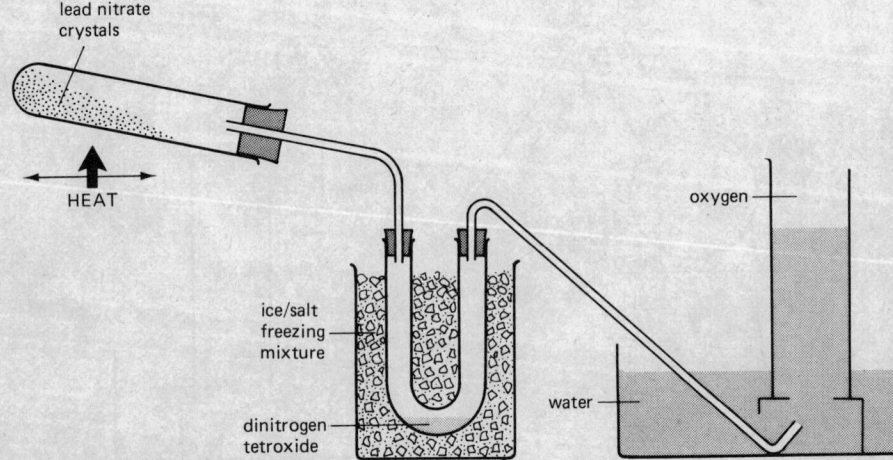

Figure 58   Preparation of dinitrogen tetroxide

N_2O_4 and NO_2. Soluble in cold water giving a mixture of nitrous and nitric acids. Hence dangerous if breathed.

Laboratory Preparation: Heating lead nitrate crystals. These are not hydrated and therefore water free (Figure 58). Crystals decrepitate (Section 19).

$$2Pb(NO_3)_2(s) \longrightarrow 2PbO(s) + 4NO_2(g) + O_2(g)$$

The NO_2 is condensed in a U-tube surrounded by a freezing mixture (ice/salt or concentrated hydrochloric acid/hydrated sodium sulphate). It condenses mainly as N_2O_4 and has only a light colour.

Industrial Preparation: Catalytic oxidation of ammonia (Section 64).

Phosphorus(III) Hydride (Phosphine), PH_3

R.M.M. = 34   R.D. = 17   Colourless. Smell of rotting fish. Insoluble. Poisonous. When impure is spontaneously flammable and preparation must be carried out in an inert atmosphere (e.g. North Sea or town gas) and in a fume chamber.

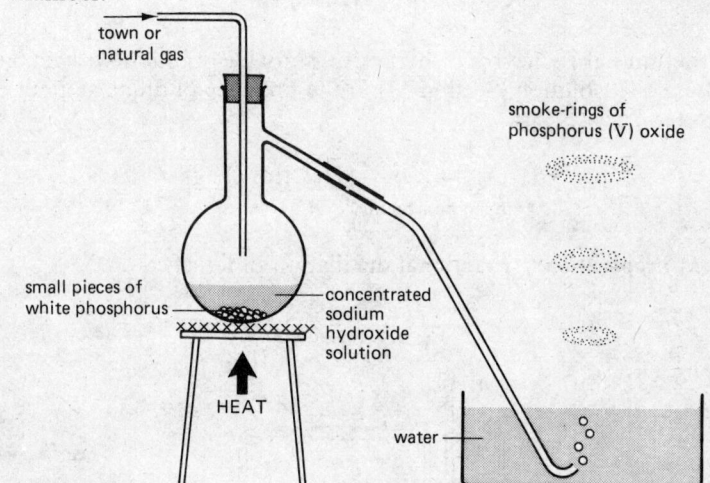

Figure 59   Preparation of phosphorus(III) hydride (to be done in a fume cupboard)

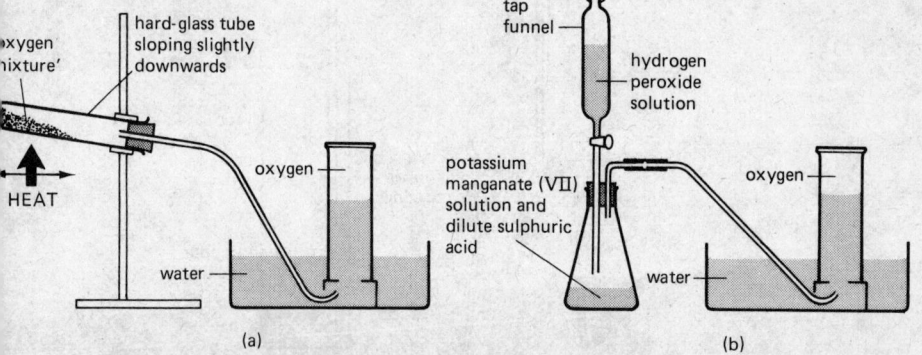

Figure 60   Preparation of oxygen

Laboratory Preparation: Warming *white* phosphorus (Section 66) with concentrated sodium hydroxide solution (Figure 59) in an inert atmosphere.

$$3NaOH(aq) + 3H_2O(l) + P_4(s) \longrightarrow 3NaH_2PO_2(aq) + PH_3(g)$$
$$\text{(sodium hypophosphite)}$$

GASES CONNECTED WITH GROUP VI ELEMENTS
(Oxygen and sulphur)

Oxygen, O$_2$

R.M.M. = 32   R.D. = 16   Colourless. Odourless. Insoluble. Neutral.

Laboratory Preparation: (1) By heating 'oxygen mixture', i.e. pure potassium chlorate mixed with *pure* manganese(IV) oxide (about one-fifth by bulk) (Figure 60(a)). The manganese(IV) oxide catalyses the decomposition of the chlorate.

$$2KClO_3(s) \longrightarrow 2KCl(s) + 3O_2(g)$$

(2) By the mutual reduction of hydrogen peroxide solution with potassium manganate(VII) solution (Section 31) in the presence of dilute sulphuric acid (Figure 60(b)).

$$H_2O_2(l) + (O) \longrightarrow H_2O(l) + O_2(g)$$
$$\text{(from manganate(VII))}$$

Industrial Preparation: Fractional distillation of liquefied air.

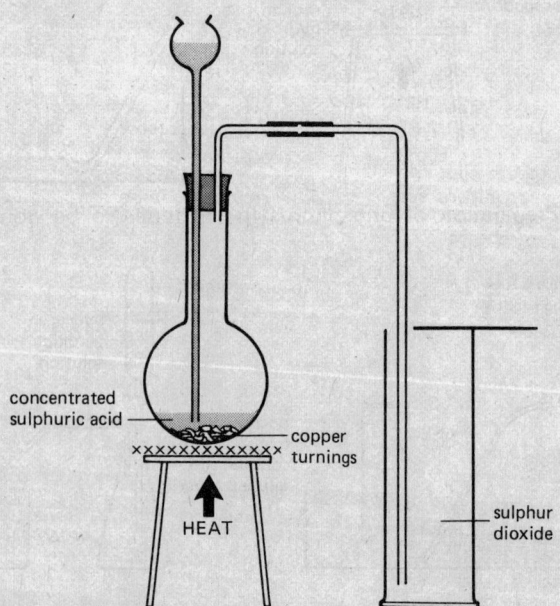

concentrated
sulphuric acid

copper
turnings

HEAT

sulphur
dioxide

Figure 61   Preparation of sulphur dioxide

Sulphur Dioxide, SO_2

R.M.M. $= 64$ R.D. $= 32$ Colourless. Choking smell of burning sulphur. Very soluble (80:1). Poisonous. Forms sulphurous acid with water which readily oxidizes to sulphuric acid.

Laboratory Preparation: Two common methods.
(1) Warming a metal (usually copper) with concentrated sulphuric acid (Figure 61).

$$Cu(s) + 2H_2SO_4(l) \longrightarrow CuSO_4(aq) + 2H_2O(l) + SO_2(g)$$

This is a complex reaction. Some insoluble sulphide is also formed.
(2) Warming a sulphite with a dilute acid.

$$SO_3{}^{2-}(aq) + 2H^+(aq) \longrightarrow SO_2(g) + H_2O(l)$$

Gas is dried through concentrated sulphuric acid.
Collected by downward delivery.

Industrial Preparation: Burning sulphur in air or oxygen: as by-product from extraction of lead and zinc: Roasting iron pyrites FeS_2 in air:

$$4FeS_2(s) + 11O_2(g) \longrightarrow 2Fe_2O_3(s) + 8SO_2(g)$$

Hydrogen Sulphide, H_2S

R.M.M. $= 34$ R.D. $= 17$ Colourless. Smell of rotten eggs. Somewhat soluble in water ($4\frac{1}{2}$:1) giving an acid solution. Poisonous. Flammable.

Laboratory Preparation: Dilute acid on a sulphide, usually iron(II) (Figure 62).

$$FeS(s) + 2HCl(aq) \longrightarrow FeCl_2(aq) + H_2S(g)$$

Kipp's apparatus can be used (Section 49, Figure 26(b)). Gas reacts with most drying agents but is usually passed through anhydrous calcium chloride. Collected by downward delivery.

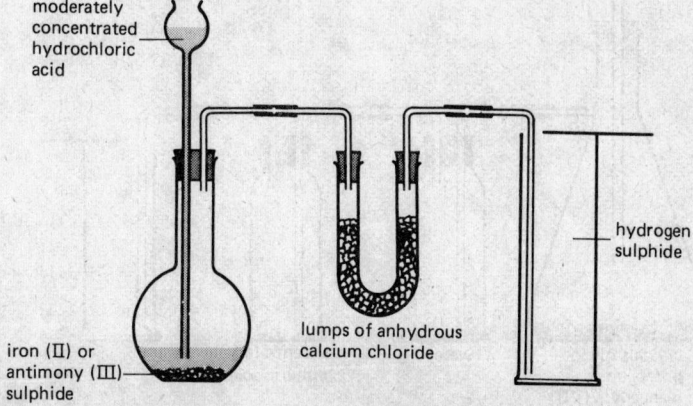

moderately concentrated hydrochloric acid

hydrogen sulphide

lumps of anhydrous calcium chloride

iron (II) or antimony (III) sulphide

Figure 62   Preparation of hydrogen sulphide (to be done in a fume cupboard)

Industrial Preparation: Not made on an industrial scale, though it does occur in some areas with natural gas and is utilized – as at Lacq, S. France.

GASES CONNECTED WITH CHLORINE–
Group VII

Chlorine, Cl$_2$

R.M.M. $= 71$ R.D. $= 35.5$ Greenish yellow. Choking smell. Somewhat soluble in water ($3\frac{1}{2}$: 1). Dissolves in cold water to give an acid solution (hypochlorous and hydrochloric). Poisonous.

Laboratory Preparation: Oxidation of hydrochloric acid (Figure 63). Most convenient agent is potassium manganate(VII). Acid is dripped slowly onto solid potassium manganate(VII).

$$2HCl(aq) + O \text{ (from oxidizing agent)} \longrightarrow H_2O(l) + Cl_2(g)$$

Dried through concentrated sulphuric acid.
Collected by downward delivery.

Industrial Preparation: Electrolysis of brine (Section 38). Oxidation of hydrogen chloride with air using a transition metal catalyst (Section 23). By-product of extraction of sodium and magnesium (Sections 39 and 41).

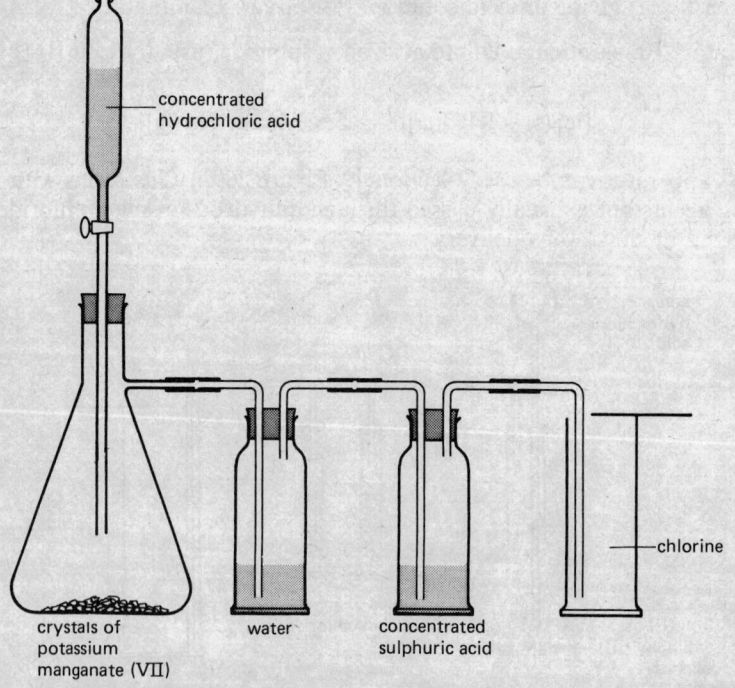

concentrated
hydrochloric acid

chlorine

crystals of
potassium
manganate (VII)

water

concentrated
sulphuric acid

Figure 63   Preparation of chlorine (to be done in a fume cupboard)

Hydrogen Chloride, HCl

R.M.M. $= 36.5$ R.D. $= 18.25$ Colourless. Choking, irritating smell. Very soluble in water (500:1) giving hydrochloric acid.

Laboratory Preparation: Warming any common chloride with concentrated sulphuric acid (Figure 64). Usually the concentrated acid is dripped onto damp rock salt (sodium chloride).

$$NaCl(s) + H_2SO_4(l) \longrightarrow NaHSO_4(s) + HCl(g)$$

Gas collected by downward delivery.

Industrial Preparation: Direct combination of hydrogen and chlorine.

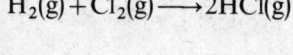

$$H_2(g) + Cl_2(g) \longrightarrow 2HCl(g)$$

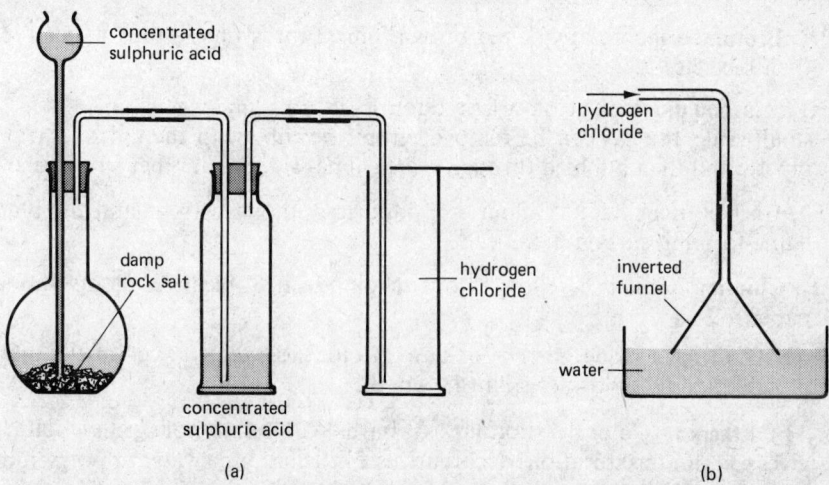

Figure 64   (a) Preparation of dry hydrogen chloride.   (b) Preparation of hydrochloric acid from hydrogen chloride

76 Simple identification of gases and vapours evolved in test-tube reactions

NH₃, Ammonia: Colourless; pungent smell; only common gas that turns moist red litmus paper blue; dense white fumes when stopper of concentrated hydrochloric acid bottle is brought near.

Br₂, Bromine vapour: Dark red/brown fumes; turns blue litmus paper red, then bleaches it.

CO₂, Carbon dioxide: Colourless; odourless; turns lime-water turbid. (On a small scale, the gas can be removed from the tube with the aid of a teat-pipette and then bubbled through a *little* lime-water in another small tube.)

CO, Carbon monoxide: Colourless; odourless; burns with a lambent blue flame forming carbon dioxide.

Cl₂, Chlorine: Greenish-yellow colour; choking smell; bleaches damp litmus paper.

N₂O, Dinitrogen oxide (nitrous oxide): Colourless; faint sweet smell; may possibly ignite a glowing splint of wood.

C₂H₄, Ethene: Colourless; odourless; burns with a luminous yellow flame; gives tests for unsaturation: decolourizes a little bromine water or potassium manganate(VII) solution.

H₂, Hydrogen: Colourless; odourless; burns with a blue flame producing water vapour which may condense on the cold side of the tube; ignites with a 'squeaky pop'.

HCl, HBr, HI, Hydrogen chloride, hydrogen bromide, hydrogen iodide: All are colourless gases that fume in moist air; dense white fumes when stopper of a bottle of ammonium hydroxide is brought near. (The bromide and iodide will nearly always be accompanied by free bromine and iodine vapour respectively.)

H₂S, Hydrogen sulphide: Colourless; smell of bad eggs; turns filter paper with lead ethanoate solution black; decolourizes potassium manganate(VII) solution and turns potassium dichromate solution green (precipitate of fine sulphur in each case).

I₂, Iodine vapour: Dense violet-coloured fumes; these condense on the cold side of the tube to a black, shiny solid.

CH₄, Methane: Colourless; odourless; burns with a blue flame producing carbon dioxide and water vapour.

NO$_2$, Nitrogen dioxide: Dark-brown fumes; turns blue litmus paper red but does not bleach it; in some cases will relight a glowing splint.

O$_2$, Oxygen: Colourless; odourless; relights a glowing splint.

SO$_2$, Sulphur dioxide: Colourless (though often has white fumes due to small quantities of trioxide); choking smell not unlike chlorine; turns potassium manganate(VII) solution colourless and potassium dichromate solution green; does not affect lead ethanoate paper.

H$_2$O, Water vapour: Colourless; odourless; condenses on the cold side of the tube and can then be tested with anhydrous copper(II) sulphate, which it turns from white to blue.

---

77   Simple tests for the identification of common ions

Notes (1) When referring to a compound, if solution is not specified, the solid, in powdered form, is to be understood. If solution *is* specified, it must be made in distilled (or deionized) water.
(2) Whenever possible, chemists carry out two tests of identification to make quite sure – a main test and a confirmatory test. In the case of a few ions, e.g. sulphate, only one good test is available.

NEGATIVE IONS (ANIONS)

CO$_3{}^{2-}$, Carbonate: A dilute acid produces characteristic effervescence and the evolution of carbon dioxide.
Heating with water has no effect (compare with hydrogen carbonate).

CH$_3$.COO$^-$, ethanoate (acetate):   Warmed with dilute hydrochloric acid gives ethanoic acid vapour (vinegary smell).
Warmed with ethanol and a few drops of concentrated sulphuric acid gives ethyl ethanoate (fruity smell).

Cl$^-$, Br$^-$, I$^-$, halide
(a) Warmed with concentrated sulphuric acid
     Nearly all *chlorides* give hydrogen chloride (fuming gas).
     Nearly all *bromides* give hydrogen bromide and bromine vapour (red/brown).
     Nearly all *iodides* give a little hydrogen iodide with iodine vapour (violet).
(b) Warmed with manganese(IV) oxide and concentrated sulphuric acid
     chloride gives chlorine gas.

bromide gives bromine vapour.
iodide gives iodine vapour.
(c) Solution with silver nitrate solution and dilute nitric acid
chloride gives a white precipitate soluble in dilute ammonia solution.
bromide gives a very pale yellow precipitate soluble only in concentrated ammonia solution
iodide gives a yellow precipitate that does not dissolve even in concentrated ammonia solution.

HCO_3^-, hydrogencarbonate (bicarbonate): A dilute acid produces characteristic effervescence and the evolution of carbon dioxide.
Heated with water also produces effervescence and carbon dioxide.

ClO^-, hypochlorite: A dilute acid produces chlorine.
Solution turns starch-iodide paper blue.

NO_3^-, nitrate
(a) Pyrolysis of most nitrates gives nitrogen dioxide and oxygen; sodium and potassium nitrates give oxygen only; ammonium nitrate gives dinitrogen oxide and steam.
(b) Warmed with concentrated sulphuric acid gives nitric acid vapour with yellow droplets of nitric acid.
(c) Warmed with copper turnings and concentrated sulphuric acid gives brown fumes of nitrogen dioxide.
(d) Solution mixed with iron(II) sulphate solution and concentrated sulphuric acid poured carefully down the side of the sloping tube – brown ring where acid meets mixture.
(e) Warmed with Devarda's alloy (containing Al, Cu, and a little Zn) and sodium hydroxide solution gives ammonia.

NO_2^-, nitrite: Warmed with dilute hydrochloric acid gives brown fumes of nitrogen dioxide.
Warmed with Devarda's alloy and sodium hydroxide solution (test (e) above) gives same result.

PO_4^{3-}, orthophosphate: Solution with silver nitrate solution gives a yellow precipitate that is soluble in ammonia solution and in dilute nitric acid.

SO_4^{2-}, sulphate: Solution with barium chloride solution and dilute hydrochloric acid (or barium nitrate solution and dilute nitric acid) gives a white precipitate of barium sulphate.

S^{2-}, sulphide: Most sulphides when warmed with dilute hydrochloric acid give hydrogen sulphide.

SO_3^{2-}, sulphite: Warmed with dilute hydrochloric acid gives sulphur dioxide.
Solution plus a little dilute sulphuric acid and warmed will decolourize a very dilute potassium manganate(VII) solution when dripped in.

$S_2O_3^{2-}$, thiosulphate: Warmed with dilute hydrochloric acid gives sulphur dioxide and a pale yellow precipitate of sulphur.

POSITIVE IONS (CATIONS)

Al^{3+}, aluminium: Solution plus sodium hydroxide solution gives a white flocculent precipitate of hydroxide that dissolves in excess of reagent. The precipitate will adsorb colouring matter to form a 'lake' (e.g. with a little litmus solution). If the hydroxide is precipitated with a solution of ammonia, the precipitate is *not* soluble in excess.

NH_4^+, ammonium: Solid or solution warmed with sodium hydroxide solution gives ammonia gas.
Many ammonium compounds sublime on heating.

Ba^{2+}, barium: Solution plus dilute sulphuric acid gives a white precipitate of barium sulphate soluble in no acid or alkali.
Flame test: Pale green

Ca^{2+}, calcium: Solution with an alkali-metal hydroxide solution only gives a white precipitate if the calcium solution is concentrated. Readily gives a white precipitate with a carbonate solution.
Flame test: Dull red

Cu^{2+}, copper(II): Solution is usually coloured blue. Solution gives the following precipitates:
With NaOH solution gives a pale-blue precipitate of hydroxide.
With NH_4OH solution gives first a pale-blue precipitate of hydroxide and then this dissolves to give the deep-blue solution containing complex ions.
With Na_2CO_3 solution gives a pale green precipitate of basic carbonate.
With H_2S gives a black precipitate of copper(II) sulphide
Flame test: Bluish-green

Fe^{2+}, iron(II): Solution usually coloured pale-green.
With NaOH solution gives a dull-green precipitate of hydroxide.
With $K_3Fe(CN)_6$ solution gives a Prussian-blue precipitate.

Fe^{3+}, iron(III): Solution usually coloured brown.
With NaOH solution gives a brown precipitate of hydroxide.
With $K_4Fe(CN)_6$ solution gives a Prussian-blue precipitate.
With KCNS solution gives a deep-red colour.

Pb^{2+}, lead(II): Solution with NaOH solution gives a white gelatinous precipitate of hydroxide soluble in excess NaOH but *not* soluble in ammonia.
Solution with Na_2CO_3 solution gives a white precipitate of basic carbonate.
Solution with H_2S gives a black precipitate of sulphide.
Solution with dilute HCl gives a white precipitate of chloride which will dissolve in hot water.

Li^+, lithium: If solution is fairly concentrated it will give white precipitates with Na_2CO_3 solution or Na_2HPO_4 solution.
Flame test: Crimson

Mg^{2+}, magnesium: Sodium hydroxide solution gives a white precipitate of hydroxide that does not dissolve in excess alkali.
Sodium carbonate gives a white precipitate of basic carbonate.

K^+, potassium: Solution gives no precipitate with any hydroxide or carbonate reagent.
Flame test: Pink/blue (lilac)

Ag^+, silver: Dilute hydrochloric acid added to the solution gives a white precipitate that is readily soluble in ammonia.
Hydrogen sulphide gives a black precipitate.

Na⁺, sodium: Solution gives no precipitate with any hydroxide or carbonate reagent.
Flame test: Bright, persistent yellow

Zn²⁺, zinc: Pyrolysis of solid. Some common zinc compounds decompose to give the oxide that is yellow when hot, white when cold. Solution plus sodium hydroxide solution gives a white flocculent precipitate that is soluble in excess of reagent. Similar reactions with ammonium hydroxide solution, but ammonium hydroxide will not precipitate the hydroxide of zinc if ammonium chloride is present.
If a precipitate forms with hydrogen sulphide, the colour of this sulphide is white.

Note: The precipitate of zinc hydroxide *is* soluble in excess ammonia solution whereas the precipitate of aluminium hydroxide *is not*.

78   Laws and problems

Law of Conservation of Matter (Lavoisier 1774): *Matter can neither be created nor destroyed in a chemical reaction.*

Also known sometimes as the *Law of Conservation of Mass* and expressed in another way: *In a chemical reaction, the total mass of the products is equal to the total mass of the reactants.*

Law of Constant Composition (Proust 1799–1802) (also called the *Law of Definite Proportions*): *All pure samples of a chemical compound contain the same elements combined together in the same proportion by mass.*

Problems

(1) **PERCENTAGE COMPOSITION BY MASS:** The formula for calcium orthophosphate is $Ca_3(PO_4)_2$. What is its percentage composition by mass? (Ca 40; P 31; O 16).

Find the relative molecular mass by adding all the relative atomic masses involved, i.e.

$$3 \times 40 + 2(31 + 4 \times 16) = 120 + 2(95) = 310$$

of this, the calcium percentage is $\dfrac{120}{310} \times 100 = 38.71\%$

the phosphorus percentage is $\dfrac{62}{310} \times 100 = 20.00\%$

the oxygen percentage is $\dfrac{128}{310} \times 100 = 41.29\%$

Total: 100%

(2) EMPIRICAL FORMULA AND CONDENSED MOLECULAR FORMULA: The percentage composition by mass of an organic compound is: carbon 60.00%; oxygen 26.67%; hydrogen 13.33%. Find its empirical (simplest) formula. If its relative molecular mass is known to be 60 and it has the properties of an alcohol what is its condensed molecular formula? (C 12; O 16; H 1)

To find the empirical formula, divide each percentage by the relative atomic mass of the element concerned:

$$\text{carbon} \quad \frac{60.00}{12} = 5.00 \quad \textbf{3}$$

$$\text{oxygen} \quad \frac{26.67}{16} = 1.67 \quad \textbf{1}$$

$$\text{hydrogen} \quad \frac{13.33}{1} = 13.33 \quad \textbf{8}$$

Find by inspection the highest common factor, which in this case is the figure for oxygen, 1.67. Divide each result by this highest common factor and the ratios are $3:1:8$, giving the simplest formula as C_3OH_8.

This corresponds to a relative molecular mass of 60 and therefore must represent the molecular formula. As it is an alcohol and has only one oxygen atom in the molecule it can have only one hydroxyl group. Therefore the condensed molecular formula must be C_3H_7OH.

Law of Multiple Proportions (Dalton 1808): *If two elements form more than one compound, then the several masses of the one which separately combine with a fixed mass of the other are in a simple ratio to one another.*

Problems

(3) Tin has two chlorides of which the percentage compositions by mass are

Chloride(i): Tin 62.63%; chlorine 37.37%
Chloride(ii): Tin 45.59%; chlorine 54.41%

Show that these figures illustrate the law of multiple proportions.

Choose one of the percentages as the 'fixed mass', say the mass of chlorine in 100 g of the first chloride -37.37 g. It is known that this mass combines with 62.63 g of tin in chloride(i). What mass of tin does this mass of chlorine combine with in chloride(ii)?

If 54.41 g of chlorine combines with 45.59 g of tin

$$37.37 \text{ g of chlorine would combine with } \frac{45.59}{54.41} \times 37.37 = 31.31 \text{ g of tin}$$

62.63 g of tin in chloride(i) and 31.31 g of tin in chloride(ii) gives the simple ratio $2:1$.

OTHER PROBLEMS INVOLVING MASS ONLY: A pure sample of anhydrous barium chloride can be made by adding excess of pure barium carbonate to dilute hydrochloric acid, warming till the reaction is finished, filtering, and evaporating the filtrate carefully to dryness.

(4) What is the maximum mass of anhydrous barium chloride that could be obtained from 9.85 g of pure barium carbonate? (Ba 137; Cl 35.5; O 16; C 12; H 1)

Write a balanced equation for the reaction:

$$BaCO_3 + 2HCl \longrightarrow BaCl_2 + H_2O + CO_2$$

Fill in the relative formula masses

$$197 \;+\; 73 \;\longrightarrow\; 208 + 18 + 44$$

(Check that the total formula masses on each side are equal)

9.85 is one twentieth of the formula mass of barium carbonate. This must therefore produce one twentieth of the formula mass of barium chloride $= \dfrac{208}{20} = 10.40\,g$ of the anhydrous salt.

(5) If the solution of barium chloride had not been evaporated to dryness but concentrated and then left for crystallization, what is the maximum mass of the hydrate $BaCl_2.2H_2O$ that could have been obtained?

In this case the formula mass would have been $208 + 36 = 244$ and the $9.85\,g$ of carbonate would have produced one twentieth of this giving $\dfrac{244}{20} = 12.20\,g$ of the hydrated chloride.

Gay-Lussac's Law of combining gaseous volumes (1809): *When gases react, the volumes in which they do so are in a simple ratio to each other and to the volume(s) of the product(s) if also gaseous, provided that all volumes are measured at the same temperature and pressure.*

Avogadro's Law (1811): *Equal volumes of gases, under the same conditions of temperature and pressure, contain the same number of molecules.*

Problems

PROBLEM INVOLVING GAS VOLUMES ONLY   (6) What is the minimum volume of air (containing one-fifth by volume of oxygen) needed for the complete combustion of $50\,cm^3$ of propane gas, C_3H_8?

First write a balanced molecular equation for the oxidation:

$$C_3H_8 + 5O_2 \longrightarrow 3CO_2 + 4H_2O$$

From this equation, 1 mole of propane needs 5 moles of oxygen. Using Avogadro's law, 1 volume of propane therefore needs 5 volumes of oxygen, or 25 volumes of air.

$50\,cm^3$ of propane needs $1250\,cm^3$ of air for complete combustion.

MOLECULAR FORMULA OF A GASEOUS COMPOUND: (7) $5\,cm^3$ of a gaseous hydrocarbon is mixed with $35\,cm^3$ of oxygen at the same temperature and pressure in a eudiometer. The mixture is sparked. After returning to the original temperature and pressure the residual gases measure $30\,cm^3$ (any steam condensing to negligible volume). After the introduction of potassium hydroxide solution, one-third of the remaining gas is absorbed, leaving

$20 \, \text{cm}^3$ of gas that proves to be oxygen. What is the molecular formula of the hydrocarbon?

The hydrocarbon can be represented as C_xH_y.
The equation for complete combustion will be (Section 53):

$$C_xH_y + \left(x + \frac{y}{4}\right)O_2 \longrightarrow xCO_2 + \frac{y}{2}(H_2O)$$

In the experiment, $35 \, \text{cm}^3$ of oxygen were introduced and $20 \, \text{cm}^3$ remained. Therefore $15 \, \text{cm}^3$ were used in the explosion. Therefore the ratio of the volume of the hydrocarbon to the volume of oxygen needed is $5:15 = 1:3$.

$$\text{Therefore } \left(x + \frac{y}{4}\right) \text{ must equal 3.}$$

Now $10 \, \text{cm}^3$ of carbon dioxide were formed. Therefore the ratio of the volume of the hydrocarbon to the volume of carbon dioxide formed is $5:10 = 1:2$.

Therefore x must equal 2 and, from the above, y must equal 4. The molecular equation therefore is:

$$C_2H_4 + 3O_2 \longrightarrow 2CO_2 + 2H_2O$$

The hydrocarbon is ethene.

Graham's Law of Diffusion of Gases (1833): *Gases diffuse at rates that are inversely proportional to the square roots of their respective densities.*

Problems:

(8) Compare the rate of diffusion of silane, SiH_4, with that of chlorine, Cl_2. (Relative density $= \frac{1}{2}$ relative molecular mass) (Cl 36; Si 28; H 1)

$$\frac{\text{Rate}_{\text{silane}}}{\text{Rate}_{\text{chlorine}}} = \frac{\sqrt{\text{density}_{\text{chlorine}}}}{\sqrt{\text{density}_{\text{silane}}}} = \frac{\sqrt{36}}{\sqrt{16}} = \frac{6}{4}$$

Silane will diffuse $1\frac{1}{2}$ times faster than chlorine.

As it is more convenient to measure times taken than relative speeds, Graham's Law can be expressed alternatively as: *The times taken by equal volumes of different gases to diffuse under the same conditions are directly proportional to the square roots of their respective densities.*

(9) If $100 \, \text{cm}^3$ of hydrogen diffuse through a porous plug in 45 seconds, how long will it take $100 \, \text{cm}^3$ of silane to diffuse through the plug under the same conditions?

$$\frac{\text{Time taken by silane}}{\text{Time taken by hydrogen}} = \frac{\sqrt{\text{density}_{\text{silane}}}}{\sqrt{\text{density}_{\text{hydrogen}}}} = \frac{\sqrt{16}}{\sqrt{1}} = \frac{4}{1}$$

Therefore $100 \, \text{cm}^3$ of silane will take $4 \times 45 \, \text{sec} = 3$ minutes.

PROBLEM INVOLVING MASS AND GAS VOLUME: (10) What volume of carbon dioxide measured at s.t.p. would be evolved if 25 g of pure calcium carbonate were treated with excess dilute hydrochloric acid? (Ca 40; O 16; C 12)

Write the balanced equation:

$$CaCO_3 + 2HCl \longrightarrow CaCl_2 + H_2O + CO_2$$

1 mole	$\longrightarrow$	1 mole
100 g	$\longrightarrow$	22.4 dm^3 at s.t.p.
Therefore 25 g	$\longrightarrow$	5.6 dm^3 at s.t.p.

If the question had asked for the volume of the gas not at s.t.p. but at 17 °C and 750 mm pressure, correction factors would have to be made using Charles' and Boyle's laws.

Correction factor for temperature (Charles' law): at 17 °C the volume will be $1\frac{17}{273}$ greater than at 0 °C. Therefore factor is: $\times \frac{290}{273}$.

Correction factor for pressure (Boyle's law): at 750 mm as the pressure is less than 760 mm the volume will be greater by the ratio $\frac{760}{750}$.

Therefore the volume at 17 °C and 750 mm will be:

$$5.6 \times \frac{290}{273} \times \frac{760}{750} = 6.028 \text{ dm}^3$$

PROBLEM INVOLVING VOLUMES OF SOLUTIONS. TITRIMETRIC ANALYSIS: A *standard solution is one whose exact concentration is known.* The concentration of a standard solution is usually expressed in grams per dm^3 (litre) of solution or in moles per dm^3 (litre) of solution. The number of moles in a dm^3 of solution is known as the *molarity* of the solution.

(11) In order to standardize (i.e. find the exact concentration of) a solution of hydrochloric acid it was titrated against an exactly molar (1 M) solution of pure sodium carbonate. 25.0 cm^3 of the acid were required exactly to neutralize 20.0 cm^3 of the carbonate solution. What was (a) the molarity of the acid; (b) its concentration in g dm^{-3}? (Cl 35.5; H 1)

Equation for the reaction:

$$Na_2CO_3 + 2HCl \longrightarrow 2NaCl + H_2O + CO_2$$

20 cm^3 of 1 M sodium carbonate solution contain 20×10^{-3} moles Na$_2$CO$_3$. From the equation, $2HCl \equiv Na_2CO_3$.

Therefore 25 cm^3 of acid contain 40×10^{-3} moles HCl.

$$1000 \text{ cm}^3 \text{ of acid contain } 40 \times \frac{1000}{25} \times 10^{-3} = 1.6 \text{ moles HCl.}$$

Therefore concentration of acid = 1.6M.
As a molar solution of hydrochloric acid contains 36.5 g dm^{-3}, its concentration in g dm^{-3} must be $1.6 \times 36.5 = 58.4$.

PROBLEM INVOLVING MASS AND VOLUME OF A STANDARD SOLUTION: Such calculations arise when considering reactions between solids and solutions. Looking again at problem (10) one might wish to know, for example, what volume of 4M hydrochloric acid would be needed to dissolve 25 g of calcium carbonate. (Ca 40; O 16; C 12)

(12) The equation $CaCO_3 + 2HCl \longrightarrow CaCl_2 + H_2O + CO_2$

shows that 1 mole of calcium carbonate requires 2 moles of the acid. In a 4M solution, 4 moles of solute are dissolved in $1 \, dm^3$ of solution. Thus 2 moles (the number required from the equation) are dissolved in $500 \, cm^3$ of the solution.

Formula mass of calcium carbonate is 100.

Therefore 100 g calcium carbonate are dissolved by $500 \, cm^3$ of 4M acid.

Therefore 25 g calcium carbonate are dissolved by $125 \, cm^3$ of 4M acid.

PROBLEM INVOLVING ELECTROLYSIS (Section 21): A number of important industrial processes depend upon electrolysis and it is necessary to know the quantity of electricity used, as for example in the refining of copper (Section 72). At the cathode, copper ions are discharged by the gain of two electrons each. The equation is:

$$Cu^{2+} + 2e^- \longrightarrow Cu$$

This shows that to produce 1 mole of copper metal, 2 moles of electrons are needed. The mass of 1 mole of electrons (6.0225×10^{23}) is so small, about 5.4×10^{-4} g, that it is not practical to attempt to weigh them. It is much easier to consider the charge carried by 1 mole of electrons. This quantity of electricity (formerly known as the Faraday) is about 96 500 coulombs. Assuming the relative atomic mass of copper to be 64, it follows from the above equation that $2 \times 96\,500 = 193\,000$ coulombs of electricity are required to liberate 64 g copper.

$$\text{Quantity of electricity} = \text{Current} \times \text{Time}$$
$$\text{(coulombs)} \qquad \text{(amperes) (seconds)}$$

(13) The time required for a given current to liberate 64 g copper, or alternatively the current required to liberate 64 g copper in a given time, can be calculated.

For example, a current of 10 A would liberate 64 g copper in $\dfrac{193\,000}{10}$ seconds – just over 5 hours 20 minutes. To produce 64 g in one hour (3600 seconds) a current of $\dfrac{19\,300}{3\,600} = 52.5$ A would be needed.

For masses other than 64 g the number of coulombs and hence the times and currents would increase or decrease in proportion to the mass. The example serves to illustrate the very high currents, with all the engineering problems that they cause, which are necessary for rapid production of materials on an industrial scale.

pre-600 B.C.   Chaldean, Egyptian, Chinese early civilizations. Practical discoveries in connection with activities such as dyeing, glass making, soap-making, embalming of bodies, perfumery, and simple metallurgy.

c. **600–0 B.C.**   The great days of Greek philosophy. Speculation and theory rather than experiment. Socrates, Plato, Aristotle, Archimedes. Beginning of atomic theory (Section 2) and ideas about elements (Section 1). Aristotle (384–322 B.C.) taught that all things must be composed of one or more of the four 'elements': earth, air, fire, and water.

c. **0–*c.* 1500 A.D.**   After the collapse of the Greek empire, such chemical knowledge as then existed passed to the Arabic races and from thence into Europe. This was the age of *alchemy*. The meaning of this word is obscure but the Arabic derivation may be the 'Black Art' (derived from the dark soil of Egypt). Arabic words, such as 'alcohol' still exist in modern chemistry. The alchemists added three more 'elements' to Aristotle's four: salt, sulphur, and mercury. During the Middle Ages, eight metals were well known: gold, silver, copper, iron, tin, lead, mercury, and antimony. The first three could be found in the native state and, with the exception of iron, the others could be extracted fairly easily. It is surprising that iron was obtained from its ores at such an early date (*c.* 1500 B.C.) as its extraction needs high temperatures and is by no means easy (Section 71). Seven of the metals were associated with the Sun, Moon, and planets and, hence, with the days of the week. Each had a symbol:

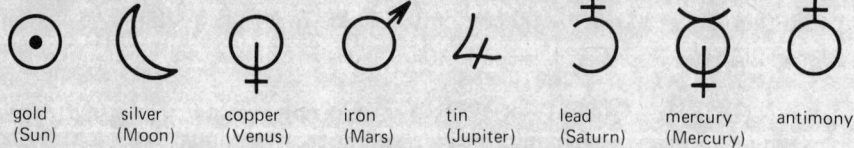

| gold (Sun) | silver (Moon) | copper (Venus) | iron (Mars) | tin (Jupiter) | lead (Saturn) | mercury (Mercury) | antimony |

Figure 65   Symbols used for different metals during the Middle Ages

These symbols probably depicted things connected with the particular gods, e.g. the shield and spear of Mars, the looking-glass of Venus, the special stave (caduceus) carried by the winged messenger Mercury. The relationship with the days of the week is most easily seen in the French language: lundi (Moonday, 'luna' being Latin for 'Moon'; mardi (Mars-day); mercredi (Mercury); jeudi (Jupiter); vendredi (Venus); samedi (Saturn). In the English language some come from the corresponding Scandinavian gods, e.g. Wednesday (Wodin's-day); Thursday (Thor's day); and Friday (Freya's day). Alchemy and astrology were much intertwined. The chief goals of the alchemist were to discover the 'philosopher's stone' which would be able to convert base metals to gold and the 'elixir of life' that would cure disease and lengthen the mortal span. There was much confused and secret symbolism. Birds, serpents, dragons, lions, and geometrical figures were all used to represent simple chemical processes. A sketch of the lion swallowing the

Sun could illustrate the dissolving of gold in 'aqua regia' (Section 73). Alchemists were familiar with simple acids and alkalis and the preparation of 'vitriols' (crystalline substances) such as the blue (copper(II) sulphate), green (iron(II) sulphate), and white (zinc sulphate). Sulphuric acid was then known as 'vitriolic acid' because it could be obtained from green vitriol. The apparatus used by alchemists included the alembic (simple distilling apparatus), crucible, phial, pestle and mortar, water bath (Bain-Marie), dung bath, ash bath, sand bath, burning glass, coal or coke furnace.

c. **1500–1700** A.D. The era of *iatrochemistry*. In this period many alchemists turned their attention to something more useful – the application of chemicals to medicine. The Greek word 'iatros' means 'physician' and we usually see it now at the end of words such as 'psychiatry'. The use of common compounds such as 'Glauber's salt' (crystalline sodium sulphate) and 'Epsom salt' (crystalline magnesium sulphate) dates from this time.

c. **1700–1800** A.D. During this century the *Phlogiston Theory* dominated men's minds. This helped to unify chemistry but impeded practical discovery. Anything that burnt was supposed to lose phlogiston in the process. Base metals when heated lost phlogiston and left the calx(ash). Phlogiston was believed to have no weight. 'Inflammable air' (hydrogen), discovered by Cavendish in 1766, was very light and considered to be almost pure phlogiston. Charcoal was also supposed to be rich in phlogiston. If a metal calx was heated with 'inflammable air' or charcoal, phlogistonists thought that the phlogiston was put back and the metal reformed. We now know that a substance burning in air *gains* oxygen.

When a metal such as zinc reacted with a dilute acid, phlogiston (hydrogen) was evolved. When the metal calx (oxide) was dissolved in acid, no phlogiston (hydrogen) was evolved. Those who supported the phlogiston theory would say that the second result was to be expected because when the metal had been heated to form the calx, the phlogiston had been driven off!

The practical discoveries that did most to kill the phlogiston theory were (1) the fact that tin and lead *gain* in weight when heated in air and (2) the discovery, towards the end of the century, of oxygen and many other gases that could not be fitted into the scheme. The great French chemist, Lavoisier, born in 1743 and guillotined by the revolutionaries in 1794, may well be regarded as the father of modern chemistry. He did away with the Aristotelian 'elements' and the phlogiston theory as well as giving us, in collaboration with other French chemists, the chemical names that we use today. All was set for a fresh start in the nineteenth century.

1800–1900 A.D. Very little of the work done in the nineteenth century has had to be discarded. Much has been extended or modified but, following the work of Lavoisier, progress in many fields was rapid. A brief list of important developments will illustrate this:

Dalton's atomic theory (Section 2)
Quantitative laws of combination (Section 78 and elsewhere)
Avogadro's hypothesis (Section 3 and elsewhere)
Determination of atomic and molecular weights
Development of organic chemistry (Section 52 and following)
Kinetic theory (Section 13)
Electrochemistry (following the work of Faraday) (Section 21)

Rise of agricultural chemistry
Industrial preparation of important chemicals on a large scale
Discovery of new elements
Classification of the elements (Sections 4, 5, and 6)
Quantitative *analysis* of mixtures and compounds
Synthesis (building up) of new compounds.

1900 onwards   The twentieth century has been marked by even more rapid progress – some dictated by two World Wars. As the subject is now too vast for one man to cover all of it, specialization is common. We have now the analytical, agricultural, physical, organic, biochemical, pharmaceutical, metallurgical, industrial, and many other types of chemist. Specially important advances have been in the study of:

Radioactive materials
Structure of the atom (Section 2)
The ways in which atoms combine (Sections 8–12)
Energy changes in chemical reaction (Section 46)
Rate of reaction and reversible reactions (Sections 47 and 48)
Development of many synthetic products such as fibres, polymers, plastics (Section 57)
Growing awareness of the evils of chemical pollution (Section 74).

In short, the subject of chemistry since the time of Lavoisier is well summed up in the *Concise Oxford Dictionary* as the 'Science of the elements and their laws of combination and behaviour under various conditions'.

Periodic Classification of

Group	I	II						Transition Serie
Period 1	1 H 1·008							
Period 2	3 Li 6·941	4 Be 9·012						
Period 3	11 Na 22·99	12 Mg 24·31						
Period 4	19 K 39·10	20 Ca 40·08	21 Sc	22 Ti	23 V	24 Cr 52·00	25 Mn 54·94	26 Fe 55·8£
Period 5	37 Rb	38 Sr 87·62	39 Y	40 Zr	41 Nb	42 Mo	43 Tc	44 Ru
Period 6	55 Cs	56 Ba 137·3	57* La	72 Hf	73 Ta	74 W	75 Re	76 Os
Period 7	87 Fr	88 Ra	89 Ac	90 Th	91 Pa	92 U 238·0	93 Np	94 Pu

* The 14 'lanthanons', elemen

nic masses (four significant figures) of the commoner elements.

				III	IV	V	VI	VII	VIII or 0
									2 He 4·003
				5 B 10·81	6 C 12·01	7 N 14·01	8 O 16·00	9 F 19·00	10 Ne 20·18
				13 Al 26·98	14 Si 28·09	15 P 30·97	16 S 32·06	17 Cl 35·45	18 Ar 39·95
o 93	28 Ni 58·70	29 Cu 63·55	30 Zn 65·38	31 Ga 69·72	32 Ge 72·59	33 As 74·92	34 Se 78·96	35 Br 79·90	36 Kr 83·80
h	46 Pd 106·4	47 Ag 107·9	48 Cd	49 In	50 Sn 118·7	51 Sb 121·8	52 Te	53 I 126·9	54 Xe 131·3
r	78 Pt 195·1	79 Au 197·0	80 Hg 200·6	81 Tl	82 Pb 207·2	83 Bi 209·0	84 Po	85 At	86 Rn
m	96 Cm	97 Bk	98 Cf	99 Es	100 Fm	101 Md	102 No	103 Lr	

inclusive, are omitted here.

Index